Jutta Metzger

Wind Systems and Energy Balance in the Dead Sea Valley

Wissenschaftliche Berichte des Instituts für Meteorologie und Klimaforschung des Karlsruher Instituts für Technologie (KIT)
Band 74

Herausgeber: Prof. Dr. Ch. Kottmeier

Institut für Meteorologie und Klimaforschung
am Karlsruher Institut für Technologie (KIT)
Kaiserstr. 12, 76128 Karlsruhe

Eine Übersicht aller bisher in dieser Schriftenreihe erschienenen Bände finden Sie am Ende des Buches.

Wind Systems and Energy Balance in the Dead Sea Valley

von
Jutta Metzger

KIT Scientific Publishing

Dissertation, Karlsruher Institut für Technologie
KIT-Fakultät für Physik

Tag der mündlichen Prüfung: 25. November 2016
Referenten: Prof. Dr. Ch. Kottmeier
PD Dr. M. Kunz

Impressum

KIT Scientific Publishing

Karlsruher Institut für Technologie (KIT)
KIT Scientific Publishing
Straße am Forum 2
D-76131 Karlsruhe

KIT Scientific Publishing is a registered trademark
of Karlsruhe Institute of Technology.
Reprint using the book cover is not allowed.

www.ksp.kit.edu

Print on Demand 2017 – Gedruckt auf FSC-zertifiziertem Papier

ISSN 0179-5619
ISBN 978-3-7315-0699-7
DOI 10.5445/KSP/1000072084

Wind Systems and Energy Balance in the Dead Sea Valley

Zur Erlangung des akademischen Grades eines

DOKTORS DER NATURWISSENSCHAFTEN

von der Fakultät für Physik des

Karlsruher Instituts für Technologie (KIT)

genehmigte

DISSERTATION

von

Dipl.-Met. Jutta Metzger

aus Pfullendorf

Tag der mündlichen Prüfung:	25.11.2016
Referent:	Prof. Dr. Ch. Kottmeier
Korreferent:	PD Dr. M. Kunz

Abstract

The Dead Sea region is endangered by severe, strongly linked environmental problems. The over-exploitation of water resources leads to a rapid lake level decline, which causes the shifting of fresh/saline groundwater interfaces, the drying up of the lake, and various climatic changes. Two key features are relevant from a meteorological point of view: the wind systems in the valley and the energy balance of the valley's surface. The wind systems influence the atmospheric conditions, such as temperature, humidity, visibility, and the air quality, while the energy balance controls the heat and moisture input to the atmosphere, and the loss of water from the water body of the Dead Sea. Therefore, it is crucial to improve the understanding of the driving forces and the characteristics of the various wind systems, including their frequency of occurrence, associated wind velocities, and their three-dimensional structure. As well, it is important to gain knowledge about the different energy balances of the main surfaces in the Dead Sea valley, i.e. water, bare soil, and vegetation. Only by having this knowledge prospective changes of the energy balance or the wind systems and the impact on other processes, such as diurnal heating of the air masses, dispersion of air pollutants in the valley, or the diurnal boundary layer development, can be assessed. Furthermore, it is important to obtain reliable data on the evaporation of the Dead Sea and to acquire knowledge about the reliability of different formulas to estimate evaporation. Evaporation is the main component of the Dead Sea water budget, which determines the rate of the lake level decline. These described challenges constitute the main topic of this thesis, and are addressed using data from measurements performed in the framework of the Helmholtz Virtual Institute DEad SEa Research VEnue

(DESERVE) (Kottmeier et al., 2016). Long-term measurements of the near-surface atmospheric conditions and the first ever performed eddy covariance measurements of the turbulent fluxes in the Dead Sea valley from March 2014 until March 2015 are combined with the data from two Special Observation Periods (SOPs). Each SOP had a duration of four weeks. The first one took place in summer and the second one in winter. The SOPs comprise comprehensive measurements of the three-dimensional structure of the atmosphere with the mobile observation platform KITcube, which combines various remote sensing and in-situ instruments.

The results of undertaken case studies, embedded into the long-term observations, reveal dominating local wind systems, which strongly influence the atmospheric conditions. During daytime, a lake breeze attenuates the temperature increase in the valley and increases the moisture content. In the evening, strong downslope windstorm-type flows (DWFs) regularly evolve triggered by the temperature difference between the air masses in the valley and at ridge height. The temperature difference is caused either by prolonged warming in the valley and radiative cooling at the ridge and at the slopes, or by the advection of cooler air masses by the Mediterranean Sea Breeze front. The DWFs are categorised according to their duration and characteristics. DWFs with a short duration of only 1 to 2 hours have a moderate mean wind velocity of 3 to 6 $\mathrm{m\,s^{-1}}$ and penetrate about 3 to 4 km into the valley. DWFs lasting the entire night influence only a shallow layer along the slopes with wind velocities similar to those of the short DWFs. The strongest DWFs last approximately 4 to 5 hours and reach a mean maximum wind velocity of more than 10 $\mathrm{m\,s^{-1}}$. They are also triggered by the temperature difference between the air masses in the valley and around ridge height but are intensified by a transition from a subcritical flow upstream to a supercritical flow downstream, as predicted by the shallow water theory. The Froude number changes from subcritical ($Fr < 1$) upstream to supercritical ($Fr > 1$) in the valley. In the case study conducted on 16 August 2014, this transition results in the formation of a hydraulic jump in the valley, a rotor beneath the hydraulic jump, and a convergence line. Such strong events lead to a replacement of the valley air mass by dry upstream air. During nighttime, northerly along-valley flows with strong wind velocities govern the atmospheric conditions, resulting in a strong latent heat flux at the water surface and a transport of the industrial emissions emitted south of the Dead Sea further to the south. The along-valley flows are triggered by a temperature gradient along the valley axis, with cooler air in the northern valley and warmer air in the south. The temperature difference is most likely caused by the differences in the cross-sectional area of the valley along the valley axis and thus a more effective cooling in the north, and by a delayed radiative cooling in the south. The delayed cooling is caused by the tempering effect of the much warmer water in the shallow evaporation ponds compared to the cooler water of the much deeper Dead Sea in the north.

Differences between the turbulent fluxes over the different surfaces are responsible for the evolution of the thermal wind systems, such as lake breeze and northerly along-valley flows, as revealed by the analysis of the energy balance measurements. Furthermore, an oasis effect over the vegetated areas is observed, resulting in considerably lower nocturnal air temperatures. A model is developed to calculate the evaporation from the Dead Sea using eddy covariance measurements and it is evaluated using a Monte-Carlo Cross Validation. The evaporation depends on the wind speed and the vapour pressure deficit. The results reveal that only the aerodynamic approach, i.e. the Dalton type equation, is able to correctly forecast the sub-daily and intra-annual cycle of the Dead Sea evaporation. Evaporation estimates calculated by using the energy balance or combination approaches are strongly biased due to the missing measurements of the heat storage term. Additionally, the hysteresis equation used to estimate the heat storage term was proven insufficient.

The results achieved prove that the atmospheric conditions in the valley are strongly coupled to the differences of the energy balance partitioning at the different surfaces and are therefore highly sensitive to the changing fraction of the land and water surface area. A further reduction of the water surface area, caused by an ongoing lake level decline, will have a strong impact on the occurrence and strength of thermally driven wind systems, on air temperature, and moisture input to the atmosphere.

Contents

1 Introduction

The Dead Sea is a unique hypersaline terminal lake and the lowest point on earth with a lake level of currently −430 m Above Mean Sea Level (AMSL) (Givati and Tal, 2016). Several important historical sites are located in the Dead Sea valley, such as Masada, Mount Sedom and the Qumran caves. Moreover, the Dead Sea valley is a well recognised centre of health treatments (Moses et al., 2006). The unique combination of the reduced UV-A radiation, the special composition of the Dead Sea water and the good air quality makes it an ideal location treating chronic skin, respiratory and rheumatic diseases (Abels and Kattan-Byron, 1985; Abels et al., 1995; Sukenik, 1996; Harari et al., 1998). In addition, large industrial operations depend on the Dead Sea and the climatic conditions in the valley. Potash, bromine, magnesium and other minerals are extracted from the Dead Sea water by means of natural evaporation. Even though the Dead Sea is of high importance to the region, it faces severe environmental challenges. Over-exploitation of the water resources of the Jordan catchment, which used to be the main surface water inflow to the Dead Sea, has reduced the inflow in the last 50 years to only 10 % of its previous amount (Salameh, 1996; Holtzman et al., 2005). Additionally, the potash companies from Israel and Jordan are pumping water from the Dead Sea into large artificial evaporation ponds for the aforementioned mineral production. Reduced inflow and the additional output of water have caused an average lake level drop of about 0.4 m yr^{-1} from 1950 until 1980 (Oroud, 2011), which increased to over 1 m yr^{-1} in the last decades (Gavrieli et al., 2005). The extreme reduction of the water body has several implications for the region. As the Dead Sea is a terminal lake, its water budget and water level influence the adjacent aquifers, such as groundwater tables, flow paths, and recharge areas (Siebert et al., 2016). As a result of the lake level decline, the groundwater tables have changed and considerable amounts of the lakebed have fallen dry (Yechieli et al., 2006). These changes result in a shift of the fresh/saline groundwater interface causing suberosion, which in turn leads to formation of sinkholes (Yechieli et al., 2006; Abelson et al., 2006). To date, over 400 sinkholes form every year, aggregating to a total of 4000 sinkholes at the western Dead Sea shoreline since the 1980s (Yechieli et al., 2016). The sinkholes affect agricultural, industrial, and environmentally protected areas, leading to substantial economic losses (Arkin and Gilat, 2000). Another consequence of the declining water level is a rapid local climate change in the region (Alpert et al., 1997). It was found that the lake breeze weakened as a result of the reduced water surface area, leading to an increase of temperature and decrease of humidity in the southern part of the valley. Additionally, the weakening of the lake breeze has led to an intensification of the Mediterranean Sea Breeze (MSB) penetration into the valley, as the lake breeze is no longer strong enough to counteract this penetration (Alpert et al., 1997). The local climate change together with the dropping groundwater levels also endanger the endemic flora and fauna. A severe dieback of vegetation and the drying up of springs, which served as a water supply for small ponds with a unique fish population, was observed in several areas around the Dead Sea, e.g. in Ein Feshkha (Goren and Ortal, 1999; Lipchin et al., 2009).

In order to gain a better understanding of the complex coupled atmospheric, hydrological and seismic hazards in this unique region, the Helmholtz virtual institute DEad SEa Rsearch VEnue (DESERVE) was founded (Kottmeier et al., 2016). DESERVE works on an interdisciplinary basis to improve the process understanding and investigate the main environmental problems, such as water budget or sinkhole formation. From an atmospheric point of view, two main components are highly relevant for the understanding of the environmental processes in the Dead Sea region. The first component includes the wind systems in the valley, which are key features of the local climate in complex terrain (Sturman et al., 1999; Whiteman, 2000). Wind systems can influence not only the atmospheric conditions, such as air temperature and humidity (Gustavsson et al., 1998; Mahrt, 2006), but also determine factors like air quality and visibility (Kalthoff et al., 2000; Darby et al., 2006). Furthermore, the wind systems strongly influence the energy balance partitioning at the surface. Wind velocity is one of the governing factors driving evaporation at a water surface (Blanken et al., 2000; Assouline and Mahrer, 1993; Nordbo et al., 2011). The energy balance partitioning at the valley's surface is the second component relevant for understanding the environmental processes in the valley. On the one hand it is strongly affected by the wind systems, but on the other hand it serves as a driver for thermally induced wind systems (Zardi and Whiteman, 2013; Vergeiner and Dreiseitl, 1987). Additionally, the energy balance links the different spheres. It influences the atmosphere, the hydrosphere, and the pedosphere (Sugita et al., 2007).

Several distinct local and mesoscale wind systems can be observed in the Dead Sea valley. Thermally driven wind systems dominate the wind field during daytime and very strong dynamically driven westerly winds are regularly observed in the evening (Bitan, 1976). The thermally induced wind systems, such as lake breeze, valley-, and slope-winds, temper the climate in the Dead Sea valley compared to the even hotter climate further to the south in the Negev desert (Shafir and Alpert, 2011). They can also be crucial for the transport and diffusion of air pollutants. Darby et al. (2006) found that especially under stable conditions, the transport of air pollutants around Salt Lake City was mainly determined by slope-wind systems. As well, Allwine (1993) showed that upslope winds can carry near-surface air pollutants to higher elevations and ventilate the valley. Dynamically driven wind systems can also affect the air quality. In arid regions, such as the Dead Sea, wind systems with high wind speeds can cause severe mineral dust emissions and can alter the transport process of mineral dust. Kishcha et al. (2016) found that strong downslope winds connected to a gust front caused strong dust emissions in the Dead Sea region, causing a concentration of total suspended particles of up to 7000 $\mu g\, m^{-3}$ in the valley. Vogel et al. (2006) found that eastward transport of mineral dust was blocked by the eastern mountain ridge of the Dead Sea valley. An additional stable stratification of the valley atmosphere can cause elevated dust concentrations in the valley itself, and dust can be transported northward along the valley axis through a channelling effect, leading to a prolonged effect on the air quality (Vogel et al., 2006). Numerous epidemiological studies have shown that an increase in PM10 concentrations is related to higher morbidity and mortality rates (Ostro, 1993; Dockery and Pope, 1994). Ostro et al. (1999) found that an increase of $10\,\mu g\,m^{-3}$ of daily PM10 was associated with an increase in the mortality rate by approximately 1 %. Thus, air quality and the correct forecast of dust emissions is of high importance for human health, particularly with regard to the treatment of chronic respiratory diseases at the Dead Sea. Dynamically driven wind systems, e.g. downslope windstorms, are relevant not only for air quality, but also have a strong impact on the greater scale wind field in the entire valley. Their high wind speeds, associated strong turbulence, and wind shear are potential threats to aviation, infrastructure, and people (Kühnlein et al. (2013) and references therein). Several extensive field campaigns have been previously conducted to investigate the four-dimensional structure of downslope winds in complex terrain. The Mesoscale Alpine Programme (MAP) investigated the driving processes and impacts of downslope windstorms on the valley atmosphere in the Alps as described in various works, e.g. by Gohm and Mayr (2004); Mayr and Armi (2008). The Terrain-induced Rotor EXperiment (TREX) focused on studying the coupling of mountain waves, rotors and the boundary layer in the Owens Valley (Grubišic et al., 2008). Several trigger mechanisms for windstorms were identified during these campaigns, including the amplification of mountain waves and density currents. Additionally, a large variety of three-dimensional structures ranging from elevated flows and Kelvin-Helmholtz waves to hydraulic jumps, rotors or convergence lines near the ground were observed. Previous to DESERVE, most investigations of the wind systems in the Dead Sea valley have been limited to data from surface stations, which limited these studies to near-surface wind conditions only (Bitan, 1974; Naor et al., 2015). The few upper air observations made in the Dead Sea valley were used

to investigate the wind channelling in wadis along the Dead Sea (Shafir et al., 2008), but not the wind field in the valley itself. To the author's knowledge, no comprehensive measurements of the three-dimensional wind field in the Dead Sea valley, which describe the specific trigger mechanisms of the wind systems, their three-dimensional structure and their impact on the valley atmosphere, have been conducted.
The driving mechanisms for local, thermally driven wind systems are the varying energy balance components at the Earth's surface. Differences in the strength of the sensible heat flux cause horizontal temperature gradients, resulting in horizontal pressure gradients, which lead to the development of wind circulations (Whiteman, 2000). An investigation of the sensible heat fluxes in the Dead Sea valley is therefore important to better understand the development of the wind systems. Besides the sensible heat flux, the latent heat flux is an important component of the energy budget of the Dead Sea. On the one hand, the latent heat flux serves as a moisture source to the atmosphere, and, on the other hand, it determines the natural loss of water from the Dead Sea (Hecht and Gertman, 2003). In arid regions, the evaporated amount of water from a lake is often large compared to other water budget components, especially to precipitation (Oroud, 2011). High uncertainties in evaporation estimates, thus, result in high uncertainties of the water budget of the lake. This makes it difficult to correctly forecast groundwater recharge and groundwater tables and, therefore, to maintain a sustainable water management (Tal, 2006; Siebert et al., 2016). Evaporation is also important for the production process of the Dead Sea Works and the Arab Potash Company, as they use the natural evaporation to gain potash and other minerals from the Dead Sea water in large artificial evaporation ponds (Ilotoviz et al., 2015). An exact forecast of daily evaporation amounts would therefore help to improve their production process considerably. Even though the evaporation rate of the Dead Sea is important, it is yet highly uncertain. Many studies have been conducted to estimate and forecast the Dead Sea evaporation using classical methods, such as the energy balance approach, water budget calculations, aerodynamic methods, or typical

combination approaches like the Penman equation (Stanhill, 1994; Oroud, 1995; Salameh, 1996; Asmar and Ergenzinger, 1999; Lensky et al., 2005; Salameh and El-Naser, 2000). These are all indirect methods, which estimate the evaporation either as a residual of the energy or water budget equations or with empirical functions. As a result, all approaches require large amounts of high quality input variables, which are often difficult to obtain. All uncertainties in the measured input variables or simplifications made to obtain the evaporation can result in large errors. Thus, current evaporation estimates for the Dead Sea vary between $1.05\,\mathrm{m\,yr^{-1}}$ (Stanhill, 1994) and $2\,\mathrm{m\,yr^{-1}}$ (Salameh and El-Naser, 1999).

All of the previously mentioned approaches have different shortcomings in estimating evaporation. Water budget methods have to account for the unknown amount of subsurface inflow to the Dead Sea (Khlaifat et al., 2010). Studies using the pan evaporation method have to consider the influence of the water activity (Salhotra et al., 1985) and studies using an aerodynamic method are often applied using an empirical wind function gained for other lakes with different properties compared to the Dead Sea (Calder and Neal, 1984). To evaluate which of these methods yields the best results for estimating Dead Sea evaporation, direct measurements of the evaporation and of the required input variables are necessary. However, to the best of the author's knowledge, no direct evaporation measurements have been performed for the Dead Sea water body so far. The only available evaporation measurements in the area are standard pan evaporation measurements conducted in the southern part of the evaporation ponds in Sedom. For estimating the evaporation of the Dead Sea the pan evaporation data are difficult to use as the heat storage of a deep lake reduces evaporation compared to the shallow pans (Morton, 1983), the activity of water in the pans is different compared to the Dead Sea water (Salhotra et al., 1985), and the local climatic conditions in Sedom are different from the conditions in the north (Alpert et al., 1997). For a thorough investigation of the Dead Sea water budget and the related processes, direct evaporation measurements are therefore imperative.

Due to the discussed impact of the wind systems on the local climate, on the energy balance partitioning, on the emission and distribution of air pollutants, and because of the possible threats caused by the high wind velocities, associated turbulence, and wind shear, the objectives of this thesis are:

(1) To determine the intra-annual variation and frequency of repeating wind systems in the Dead Sea valley, to assess their characteristics and their three-dimensional structure, and to identify particular trigger mechanisms for the evolution of the wind systems.

(2) To quantify the different energy balance partitioning at the main surfaces in the valley, i.e. bare soil, water, and vegetation, and to determine the influence of the energy balance on the wind field and on the local climatic conditions.

Moreover, it was discussed that the energy balance not only interacts with the wind field and influences the atmospheric conditions, but is also crucial for the water budget of the Dead Sea, the hydrological system, and for industrial operations. Therefore, further objectives are:

(3) To determine the actual evaporation rate of the Dead Sea water, its intra-annual variability, and its dependence on meteorological conditions.

(4) To find a model that satisfyingly estimates the turbulent heat fluxes of the Dead Sea water surface and to investigate the reliability of commonly used evaporation formulas on different time scales.

Finally, the feedback mechanisms between the wind systems and the energy balance components will be discussed, with regard to a continues lake level decline in the future.
To address these scientific questions, a comprehensive database is necessary. Therefore, a long-term monitoring network was established in the framework of DESERVE. It included long-term measurements of the near-surface atmospheric conditions and eddy covariance measurements performed for the first time in

the Dead Sea valley, which were simultaneously performed over bare soil, vegetation, and Dead Sea water. Additionally, comprehensive four-dimensional measurements of the atmospheric conditions were conducted during two Special Observation Periods (SOPs) with the mobile observation platform KITcube, which combines in-situ and remote sensing instruments to capture multi-scale atmospheric processes (Kalthoff et al., 2013). The long-term measurements, the energy balance measurements, and measurements from the SOPs were combined to address the scientific objectives.
In the following Chap. 2 the theoretical background of the wind systems and the energy balance is presented. In Chap. 3 the research area and the measurements are described. Methods used for data post-processing and complementing measurement results necessary for the postprocessing are described in Chap. 4. An analysis of the observed wind systems, their characteristics, and their driving processes in Chap. 5 are followed by the analysis of the energy balance partitioning at the valley's surface, the presentation of a regression model for the turbulent fluxes of the water surface, and an evaluation of commonly used evaporation formulas in Chap. 6. In Chap. 7, a summary and an explanation of the interaction between wind systems and energy balance components is given, followed by the overall conclusions in Chap. 8.

2 Phenomenological and Theoretical Background

This thesis focuses on wind systems and surface energy balances and thus on two important atmospheric topics which are highly relevant for the understanding of processes threatening the Dead Sea area, e.g. the lake level decline, transport of air pollutants, and climatic changes. To understand the experimental results described in Chap. 5 and 6 the theoretical background for the topics will be explained in this chapter. First, thermally driven wind systems and their possible superposition are described in Sec. 2.1. In Sec. 2.2 the influence of large-scale flow on the Atmospheric Boundary Layer (ABL) in valleys is presented, and in Sec. 2.3 the energy balance partitioning at the Earth's surface, together with a direct method to measure the turbulent fluxes and empirical formulas to estimate them is discussed.

2.1 Thermally Driven Wind Systems

Under weak synoptic forcing and fair weather conditions thermally driven wind systems can develop. These wind systems are caused by differential heating of atmospheric layers either above different land surfaces or along slopes. The horizontal temperature differences result in pressure gradients, which cause a flow from areas with low temperatures and thus higher pressure to warmer areas with lower pressure near the surface. Above a return flow leads to a closed circulation (Zardi and Whiteman, 2013). Due to the Coriolis force the direct circulation adjusts into the direction of the geostrophic flow after a certain amount of time (Kottmeier et al., 2000). The relevant time scale is

$$\tau_f = \frac{2\pi}{f_c}, \tag{2.1}$$

with the Coriolis parameter, $f_c \approx 0.75 \cdot 10^{-4}\,\mathrm{s}^{-1}$, for the Dead Sea area. The vertical change of the geostrophic wind is called thermal wind and is oriented parallel to the isotherms.
Thermally driven wind systems are often characterised by a wind direction reversal twice per day. Their spatial scale, strength, duration, depth and onset time vary strongly and are mainly influenced by the energy balance of the surface, the large-scale wind conditions and terrain characteristics. Typical thermally driven wind systems within complex terrain are slope-wind systems and valley-wind systems. Sea- or lake-breezes are typical wind systems induced over adjacent water and land surfaces.

2.1.1 The Sea-Breeze Circulation

Sea-breeze circulations can be observed at all sea coasts or at the shore of larger lakes. Through the higher specific heat capacity of the water and a higher molecular heat conductivity compared to soil, the sensible heat flux at the water surface is much smaller than at the land surface and thus the near-surface air temperature over the land increases faster than over the water (Stull, 1988). Thereby, the air temperature over the land has a much larger diurnal variation than the air over the water surface. This results in a horizontal temperature and pressure gradient perpendicular to the shoreline and a baroclinic atmosphere. During daytime a flow along the horizontal pressure gradient from the higher pressure over the water surface towards the lower pressure over the land surface develops: the sea-breeze. During nighttime the pressure and temperature gradients are reversed and a land-breeze develops, which is much weaker than the sea-breeze (Simpson, 1994). The landward edge of the sea-breeze is marked by strong temperature and humidity gradients between the moist and cool maritime air and the warmer and drier air over land. Therefore, it is often called the sea-breeze front. Typical maximum wind velocities observed for sea-breezes vary from 2 up to 11 m s^{-1} (Atkinson, 1981; Defant, 1950) and typical distances of its inland penetration in mid latitudes vary from only 15 km observed by Wexler (1946) in the UK up to 150 km as reported by Kottmeier et al. (2000) at the coast of Spain. The vertical extent of the sea-breeze varies between 100 and 1000 m according to Atkinson (1981). The sea-breeze characteristics, like inland penetration, depth, vertical and horizontal wind velocity, depend on several factors. A detailed review and discussion of the dependencies of the sea-breeze characteristics on various influencing factors and possible scaling laws provide Crosman and Horel (2010). Here, only a summary of their review shall be given. Particularly the sensible heat flux at the surface, which determines the strength of the temperature gradient, but also atmospheric stability and the geostrophic wind influence the sea-breeze characteristics. With increasing sensible heat flux at the land surface, the inland penetration, the depth, the horizontal and vertical wind velocities of the sea-breeze increase (Crosman and Horel (2010) and references therein). Increasing stability on the other hand, reduces the depth and the vertical wind velocities, but does only slightly influence the inland penetration and the horizontal wind velocity (Porson et al., 2007a; Antonelli and Rotunno, 2007; Crosman and Horel, 2010). The influence of the geostrophic wind depends on its wind direction and wind speed. Moderate offshore winds relocate the sea-breeze front towards the shoreline and strengthen the sea-breeze front gradients of humidity and temperature. Onshore flows lead to a suppression of the frontogenesis and a weakening of the temperature gradient of the sea-breeze front and therefore weakens the sea-breeze circulation (Arritt, 1993; Porson et al., 2007a). Savijärvi and Alestalo (1988) and Arritt (1993) reported that for an onshore geostrophic flow of about 2 to 4 m s^{-1} the sea-breeze is no longer distinguishable from the background flow.

2.1.2 The Slope-Wind Circulation

Slope flows can either be upslope winds during daytime or downslope winds during nighttime. Through a positive sensible heat flux during daytime at the slopes, the air near the slope is heated and is thus warmer than air at the same height over the valley (Stull, 1988). Equally during night the air near the slope is radiatively cooled and thus colder than the air at the same height in the valley. The main driving mechanism of a slope flow is buoyancy, which is positive during the day and negative during the night. Temperature perturbations along the slope can additionally contribute to the slope flows. Increasing temperature perturbations upslope lead to small upslope wind contributions, whereas decreasing temperature perturbations upslope lead to a reduction of the upslope flow (Haiden, 2003). Typical wind speeds which are reached for upslope winds are 1 to 5 $m\,s^{-1}$ and the depth of the upslope flow layer is between 20 to 200 m Above Ground Level (AGL). For downslope flows 1 to 4 $m\,s^{-1}$ are reached and a depth of 3 to 100 m AGL (Zardi and Whiteman, 2013). They both have a jet like structure with a wind maximum near the ground at about 1 to 15 m for downslope flows and 10 to 50 m for upslope flows (Zardi and Whiteman, 2013). Through this jet like structure of the downslope flow, turbulent kinetic energy (TKE) shows a maximum below the wind speed maximum, a local minimum at the height of the jet maximum, and TKE above the wind speed maximum depends on the wind in the layer above the downslope layer. Thus, turbulence above the wind speed maximum is decoupled from the surface (Horst and Doran, 1988). Those slope flows are non-stationary as they correspond directly to changes of the energy balance of the surface. The onset of the upslope winds is just some minutes after sunrise when the sensible heat flux gets positive, and the upslope winds increase in strength as long as the sensible heat flux increases. On the other hand, upslope winds can stop immediately, e.g. when clouds cover the sun and thus solar radiation decreases suddenly (Vergeiner and Dreiseitl, 1987). Downslope flows start immediately after sunset when the radiative cooling of the near-surface air mass starts.

2.1.3 The Valley-Wind Circulation

Valley-wind circulations develop due to a horizontal temperature gradient along the valley axis or between the valley and the adjacent plains. During daytime, when the air in the valley is warmer than further down in the valley or over the plain, a horizontal pressure gradient evolves and causes a near-surface up-valley wind. During nighttime the temperature gradient is reversed leading to a down-valley wind (Whiteman, 2000). The onset of the valley-winds is several hours after the onset of the up- and downslope winds because the whole valley air mass has first to be heated or cooled before the valley wind sets in (Zardi and Whiteman, 2013). Typical wind speeds of the near-surface winds are in the range of 3 to $10\,\mathrm{m\,s^{-1}}$. The upper branch of the circulation runs in opposite direction, but is often not confined by topography and thus broader in its horizontal extent, leading to weaker wind speeds of about 1 to $2\,\mathrm{m\,s^{-1}}$ or it is obscured by the large-scale flow (Zardi and Whiteman, 2013). The stronger heating of the valley air mass during daytime and the stronger cooling during nighttime compared to the adjacent plains has mainly two factors. Firstly, the valley atmosphere is often decoupled from the large-scale flow, so that no air mass exchange takes place with the large-scale flow. The radiative heating or cooling is concentrated on the air mass in the valley and additional radiative warming or cooling from the slopes makes it even more effective (Zardi and Whiteman, 2013). The second factor is called the Topographic Amplification Factor (TAF), which is probably more important. This concept was first expressed by Wagner (1932) and further investigated by Steinacker (1984). The warming of the valley atmosphere is stronger compared to the warming of a volume, which has the same depth and area at the top of the volume over the plains, because the received solar radiation, which is available for heating the air, is the same but the air volume which has to be heated is smaller in the valley through the confining slopes. This concept can also explain along-valley wind speed variations, caused by along-valley pressure perturbations, as the terrain cross-section varies along the valley axis (McKee and O'Neal, 1989). However, applying this concept in practise also reveals several difficulties, especially in defining the top of the volume, the heat fluxes across the top boundary of the volume through slope-winds or subsidence (Schmidli and Rotunno, 2010; Rampanelli and Zardi, 2004; Weigel et al., 2007), and the influence of tributary valleys (Steinacker, 1984).

2.1.4 Superposition of Thermally Driven Wind Systems

Slope and Valley Winds

The slope and valley wind systems described in Chap. 2.1.2 and 2.1.3 normally superimpose during the day. The superposition can be divided into four phases after Zardi and Whiteman (2013).

The morning transition phase starts shortly after sunrise, when the sensible heat flux gets positive. Through the heating at the sidewalls, upslope winds develop. The morning transition ends with the break-up of the nocturnal boundary layer inversion. The valley atmosphere is heated by two processes. The growth of a convective boundary layer and the subsidence in the middle of the valley to compensate for the upslope winds (Zardi and Whiteman, 2013). When the air in the valley is heated sufficiently, up-valley winds evolve.

After the break-up of the nocturnal boundary layer inversion convection grows and can even extend above ridge height, making a coupling to the large-scale flow possible. This is called the daytime phaes. If the along valley pressure gradient is still maintained, the up-valley flow continues. This is possible through the TAF which still contributes to a stronger heating of the valley air mass than over the plains (Zardi and Whiteman, 2013). They daytime phase normally lasts from mid morning until late afternoon.

The evening transition phase starts when the energy balance reverses and the sensible heat flux gets negative. The near-surface air starts to cool and downslope winds develop. Those downslope winds converge and in the valley centre the cool air mass rises, cooling the entire valley cross-section (Whiteman, 2000; Brazel et al., 2005). The up-valley wind still continues to blow as long as the air mass in the valley is not cooled sufficiently to be colder than the air masses over

the plains. Several hours after sunset the pressure gradient finally reverses and an down-valley wind evolves. The superposition of the downslope flows and the down-valley winds is often referred to as a drainage flow.
The nighttime phase is mostly defined by a stable stratification and a dominating down-valley flow. The nighttime phase starts several hours after sunset and lasts until sunrise.

Sea-Breeze and Mountain Winds

Over mountains near a shoreline, a superposition of the sea-breeze and the slope- and valley-winds can occur. Mountains which are aligned parallel to the shoreline can either enhance or suppress the sea-breeze. A very steep slope can mechanically block the sea-breeze front (Ramis and Romero, 1995; Barthlott and Kirshbaum, 2013), but if the slope is sufficiently gentle the sea-breeze can couple to the upslope winds when the diurnal temperature variations are in phase (Miller, 2003). This amplifies wind speeds as well as the depth of the sea-breeze layer and its inland penetration (Mahrer and Pielke, 1977; Porson et al., 2007b). Coastal valleys generally increase the wind speeds and the inland penetration of the sea-breeze front as the sea-breeze is channelled along the valley axis (Miller, 2003).

2.2 Impact of Large-Scale Flows on the Valley ABL

The large-scale flow can interact with the valley ABL through various processes. Mechanically induced turbulence at the valley ABL top can lead to a downward mixing of the upper air masses. Dynamically driven wind systems, e.g. mountain waves or downslope windstorms, can strongly impact the valley ABL, or large-scale induced wind channelling along the valley axis can occur. The extent of the influence depends on the stratification of the atmosphere and on the wind speed and direction of the large-scale flow (Whiteman, 2000). Especially when the large-scale forcing is strong or the inversion at the valley boundary layer top is weak, the impact on the conditions in the valley can be large.

2.2.1 Turbulent Mixing

When the large-scale flow is much stronger than the valley wind systems, strong vertical wind shear can cause turbulence which erodes the valley ABL inversion. Humidity, momentum, and heat can be transported downward into the valley and overpower the thermally driven wind systems. Wind speeds of at least 6 to $7\,\mathrm{m\,s^{-1}}$ are necessary to erode a medium to strong inversion (Whiteman, 2000). Such wind break-ins are most likely under unstable or neutrally stratified conditions (Whiteman and Doran, 1993). To evaluate whether turbulence is produced by buoyancy or is mechanically induced by wind shear the gradient Richardson number Ri is often used as a criterion. It describes the ratio between buoyancy produced and mechanically induced turbulence and can be calculated with

$$Ri(z) = \frac{g}{\overline{\Theta}(z)} \frac{\frac{\partial \overline{\Theta}(z)}{\partial z}}{\left(\frac{\partial \overline{u}(z)}{\partial z}\right)^2 + \left(\frac{\partial \overline{v}(z)}{\partial z}\right)^2}, \tag{2.2}$$

e.g. after Stull (1988), with the mean potential temperature profile, $\overline{\Theta}(z)$, the gravity acceleration, g, and the profiles of the mean horizontal wind components, $\overline{u}(z)$ and $\overline{v}(z)$. Values of $Ri < 0$ indicate an unstable stratification and turbulence is produced through buoyancy and through wind shear. In a stably stratified atmosphere turbulence is produced mechanically but at the same time it is dissipated through negative buoyancy. Only when the values of the Richardson number are smaller than a critical value of $Ri_c = 0.25$ mechanically produced turbulence overpowers the negative buoyancy and a stably stratified layer can become turbulent (Etling, 2008). In this case unstable Kelvin-Helmholtz waves may develop. If $Ri < Ri_c$ in the inversion at the valley ABL top Kelvin-Helmholtz waves develop in this layer and keep growing until they break, leading to an unstable layering of air with higher density above air with lower density. This causes buoyancy and thus turbulent mixing of the warmer drier air from above with the cooler, moister air in the boundary layer, resulting in a reduction of sharp temperature and moisture gradients at the valley ABL top. The temperature inversion is shifted to lower levels.

2.2.2 Dynamically Driven Flows

Mountains can have a large influence on the wind field. An air mass can either be blocked and forced to flow around the mountain, it can flow over the mountain or it can pass through gaps. The non-dimensional mountain height, $\hat{H}$, provides information about the flow conditions (Jackson et al., 2013). It is defined as

$$\hat{H} = \frac{Nh_c}{u_{up}}, \tag{2.3}$$

with the crest height, h_c, the undisturbed horizontal wind speed upstream, u_{up}, and the Brunt-Vaisala frequency, N, which is defined as

$$N = \sqrt{\frac{g}{\overline{\Theta}} \frac{\partial \Theta}{\partial z}}, \tag{2.4}$$

with the potential temperature, $\overline{\Theta}$, and height z. $\hat{H}$ is normally calculated for a layer, and thus $\overline{\Theta}$ and u_{up} represent mean values of this layer. For a non-dimensional mountain height $\hat{H} \leq 1$ the flow has sufficient kinetic energy to flow over the mountain, but for low wind speeds or strongly stable stratification the air mass is blocked by the mountain indicated by $\hat{H} > 1$. When the flow is not blocked by the mountains and flows over the mountain, it can influence the downstream atmosphere. An important theory, with which such dynamically driven flows over mountains can be described, is the hydraulic flow theory first presented by Prandtl et al. (1942).

To give a physical explanation of a dynamically driven flow over a mountain, it is assumed that air flowing over a mountain is similar to thin water layer flowing over an obstacle, e.g. a rock. The basic concept will be explained using the one-dimensional case of a flowing layer with a free surface at height $\eta(x)$ flowing over a terrain $h(x)$. Using the shallow water equations, meaning that the horizontal length scale is much greater than the vertical length scale, and assuming a frictionless steady state flow without considering the Coriolis force or external pressure gradients the momentum equation and the continuity equation can be written to (e.g. Jackson et al. (2013))

$$u\frac{\partial u}{\partial x}+g\frac{\partial \eta}{\partial x}=g\frac{\partial h}{\partial x} \tag{2.5}$$

$$\frac{\partial (u\eta)}{\partial x}=0, \tag{2.6}$$

where u is the speed of the flowing layer in x-direction. Eqs. 2.5 and 2.6 can be combined to

$$\left(1-\frac{u^2}{g\eta}\right)\frac{\partial \eta}{\partial x}=-\frac{\partial h}{\partial x}. \tag{2.7}$$

The shallow water phase speed is

$$c\equiv\sqrt{g\eta} \tag{2.8}$$

and its relation to the flowing layer speed u results in the Froude number

$$Fr=\frac{u}{\sqrt{g\eta}}. \tag{2.9}$$

Eq. 2.7 can then be rewritten to

$$\frac{\partial \eta}{\partial x}=-\left(\frac{1}{1-Fr^2}\right)\frac{\partial h}{\partial x}. \tag{2.10}$$

The characteristics of the flow can be explained by the Froude number. In a subcritical case ($Fr< 1$) waves can travel up- and downstream. When the flow is supercritical ($Fr > 1$) waves can only propagate downstream. $Fr = 1$ is called the critical state. Three situations can occur when a flow passes an obstacle. Firstly, the layer is subcritical everywhere. Then the layer height has to decrease when the terrain rises and it has the shallowest depth at the crest. Secondly, the flow is supercritical everywhere, then the layer height and the terrain height rise simultaneously. Thirdly, the layer changes from a subcritical state upstream to supercritical state downstream. This means that the layer has to be thick and slow upstream and fast and shallow downstream. Further downstream the layer eventually thickens and slows down in a hydraulic jump. This causes strong turbulence as the momentum is conserved but kinetic energy dissipated (Jackson et al., 2013). For the transition of the state from subcritical to supercritical, the flowing layer has to reach its critical state $Fr = 1$ at crest height. The Bernoulli equation, describing the conservation of energy is then

$$\frac{1}{2}u^2 + g\eta_c = gh_{up}, \tag{2.11}$$

where h_{up} is the layer height far upstream, resulting in an increase of the layer speed at crest height of $u_c = \frac{3}{2}u_{up}$ compared to the conditions upstream and the layer depth has to decrease by $\eta_c = \frac{2}{3}h_{up}$ (Jackson et al., 2013). Even though the hydraulic flow theory can describe processes in the atmosphere the application of the theory in practise yields several difficulties. In a continuously stratified medium, such as the atmosphere, no free surfaces exist (Durran, 2003). In the real atmosphere, wind speed and stratification are often not constant with height and different layers are only separated by indistinct inversion. This makes the calculation of the Froude number quite difficult. However, if there is a strong well-defined inversion at height, h, it can be used to define two layers with constant potential temperature and wind speed. The Froude number can be calculated using a 'reduced gravity'

$$g' = g\frac{\Delta\Theta}{\Theta_m}, \tag{2.12}$$

where $\Delta\Theta$ is the inversion strength and Θ_m is the mean potential temperature of the lower layer, resulting in

$$Fr = \frac{u}{\sqrt{hg'}}. \tag{2.13}$$

Downslope Windstorms

Downslope windstorms may be very strong and often gusty, blowing down the lee slope of a mountain range. They are often named differently depending on the geographic region of their occurrence. Foehn is one of these local names, referring to a warm and dry downslope wind with high wind speeds in the Alps. Several mechanisms were proposed to describe the initiation of downslope windstorms. The amplification of vertically propagating gravity waves by an interference at a critical layer was suggested by Klemp and Lilly (1975), the reflection of propagating waves at a self-induced critical layer, caused by wave breaking (Peltier and Clark, 1979) and by means of the hydraulic flow theory as a transition of a sub- to supercritical flow (Long, 1954; Durran, 1986; Klemp and Durran, 1987). The hydraulic flow theory was recently used in several large field campaigns such as the Mesoscale Alpine Programme (MAP; Mayr and Armi (2008)) or the Terrain-Induced Rotor Experiment (TREX; Jiang and Doyle (2008), Armi and Mayr (2011)) to explain downslope windstorms triggered by a density current. Jiang and Doyle (2008) and Mayr and Armi (2010) found that a negative potential temperature difference between the air in the valley and the air at crest height is necessary for the air to descent into the valley. Are the wind velocities of these density currents not adequate to use the term 'storm', they are also called downslope windstorm-type flows (DWFs), e.g. (Lehner et al., 2016). With this expression they are differentiated from the normal downslope winds, and at the same time, indicate that they have similar characteristics and trigger mechanisms, as downslope windstorms.

2.2.3 Wind Channelling

Two types of wind channelling in valleys can be distinguished, forced channelling or pressure-driven channelling. Forced channelling describes the situation when the large-scale flow is aligned with the valley axis through the channelling by the valley sidewalls. The channelled flow is a projection of the along valley component of the large-scale flow, and the wind direction (up- or down-valley) depends on the relative angle of the large-scale flow to the valley axis (Whiteman and Doran, 1993). The wind is always channelled in the direction that requires the smallest directional change compared to the large-scale flow (Whiteman, 2000). This forced channelling often affects the wind field in the upper part of a valley near the ridge. Fiedler (1983) proposed another trigger mechanism for along-valley flows, which is referred to as pressure-driven channelling. The channelling is caused by the large-scale pressure gradient along the valley axis and can either blow in the same direction as the along-valley wind component of the large-scale flow, or in the opposite direction. This phenomenon is then called counter-current (Kalthoff and Vogel, 1992; Wippermann, 1984). Whiteman and Doran (1993) described the along-valley pressure gradient with

$$\frac{\partial p}{\partial x} = |\nabla p| \sin\alpha. \tag{2.14}$$

The positive x-axis is oriented in the up-valley direction and α is defined as the counterclockwise angle between the positive x-axis and the geostrophic wind. With this definition an up-valley wind evolves when the pressure gradient is negative whereas a down-valley wind evolves when the pressure gradient is positive. The strongest channelling occurs when the geostrophic wind is perpendicular to the valley axis (α =90° or 270°). No pressure-driven channelling occurs for angles close to 0° or 180°. Counter-currents occur when α is between 0° and 90° or between 180° and 270°. In the vertical, wind will turn clockwise with increasing height above the valley floor.

2.3 Energy Balance of the Earth's Surface

The main energy transformation for atmospheric processes takes place at the Earth's surface. The main energy input is the solar radiation flux ($K\downarrow$). Depending on the surface properties, part of the solar radiation flux is reflected at the surface, which is expressed by the albedo (α). Another energy input at the

Earth's surfaces is the longwave downward radiation ($L\downarrow$), which is emitted by clouds, aerosols or gaseous components of the atmosphere. The emitted energy is proportional to the fourth power of their thermodynamic temperature (T_B) and can be expressed with the Stephan-Boltzmann law

$$L\downarrow = \varepsilon \cdot k_B \cdot T_B^4, \qquad (2.15)$$

with the Stephan-Boltzmann constant, k_B, and the emissivity coefficient, ε. Similarly to aerosols or clouds, the earth also emits energy ($L\uparrow$) proportional to its surface temperature (T_s). This energy can be calculated using Eq. 2.15. The net radiation at the Earth's surface is the sum of those four radiation terms

$$Rn = (1-\alpha) \cdot K\downarrow + L\downarrow - L\uparrow, \qquad (2.16)$$

and results in a surplus of energy during the day and an energy deficit during the night (Foken, 2006). These diurnal deviations are balanced by the sensible heat flux (H), latent heat flux (LE), and the ground heat flux (G) as follows

$$Rn = H + LE + G. \qquad (2.17)$$

Net radiation is defined as positive when it is directed towards the surface and negative when it points away from the surface, and the other terms are defined the opposite way.

The energy balance equation is only valid for surfaces. For the investigation of a certain volume, such as a water body, the heat capacity and the possibility of advection of energy into the volume have to be taken into consideration. The sum of the advected heat fluxes (F_n) due to water inflow, water outflow, and precipitation, as well as a storage term (ΔQ) have to be added to Eq. 2.17. The general form of the energy budget equation of a water body is

$$Rn = H + LE + G + F_n + \Delta Q. \qquad (2.18)$$

The ground heat flux of the energy budget equation represents the heat flux between the water and the soil at the bottom of the lake.
The ground heat flux is mainly a molecular heat transport and is proportional to the soil temperature gradient multiplied by the molecular heat transfer coefficient (Stull, 1988). The sensible heat flux describes the heating of the near-surface air masses through turbulent heat transfer along the vertical temperature gradient, and the latent heat flux describes the water vapour flux and the energy stored in the phase transition (Foken, 2006). The equations are

$$G = a_G \frac{\partial T_{soil}}{\partial z}, \tag{2.19}$$

$$H = -c_p \cdot D_H \frac{\partial T_a}{\partial z}, \tag{2.20}$$

$$LE = -L_v \cdot D_{LE} \frac{\partial q}{\partial z}, \tag{2.21}$$

with the molecular heat transfer coefficient of the soil, a_G, soil temperature, T_{soil}, specific heat capacity at constant pressure, c_p, air temperature, T_a, latent heat of vaporisation, L_v, specific humidity, q, and the turbulent transfer coefficients D_H and D_{LE}. These coefficients depend on surface properties, wind speed, and stability (Foken, 2006). The latent heat of vaporisation of freshwater depends on the water temperature (T_w) and can be described by

$$L_v = 2501 - 2.37 \cdot T_w, \tag{2.22}$$

with L_v in in $\mathrm{kJ\,kg^{-1}}$ and T_w in K.

2.3.1 Eddy Covariance Method

There are several methods to assess the turbulent components of the energy balance, e.g. the Bowen ratio method or bulk approaches, but the only method to directly measure the fluxes is the eddy covariance method. The eddy covariance method describes the heat, mass and momentum transfer between a surface and the atmosphere as the covariance of the turbulent fluctuations of the vertical wind and the desired quantity (Foken et al., 2012a). This method, first proposed by Montgomery (1948), Swinbank (1951) and Obukhov (1951) is based on the mathematical description of the temporal change of a quantity in a turbulent boundary layer, with several assumptions and simplifications. A detailed description and full derivation of the equations can be found in most standard textbooks about micrometeorology, e.g. Stull (1988), Foken (2006) and will therefore only be summarised following Foken et al. (2012a). The temporal change of any quantity in the atmosphere can be described by atmospheric transport, molecular diffusion, and a source or sink of the quantity in the defined volume. As in the atmospheric boundary layer turbulent transport is the dominant transport mechanism, the Reynolds decomposition and the Reynolds postulates, describing turbulent motions by decomposing the variables in a mean part and a fluctuation, have to be applied. This approach requires statistical stationarity of the fluctuations during the averaging period. Assuming horizontal homogeneity and stationarity, removes all time derivatives and horizontal gradients, such that

$$H = \rho_a \cdot c_p \cdot \overline{w'T'}, \tag{2.23}$$

$$LE = \rho_a \cdot L_v \cdot \overline{w'q'}. \tag{2.24}$$

ρ_a is air density and w', q', and T' are the deviations from the mean vertical wind speed, the mean specific humidity, and the mean temperature. Evaporation (Ev) is then calculated from the latent heat flux using following equation

$$Ev = \frac{LE}{\rho_w \cdot L_v}. \tag{2.25}$$

2.3.2 Empirical Evaporation Formulas

To estimate evaporation from open water, several formulas have been proposed in the past. They can be grouped by the approach applied to get the formula, including the energy budget equations, the aerodynamic approaches, also known as Dalton type formulas, and the combination approaches. Overviews and comparisons of the different equations can be found in papers or books (e.g. Drexler et al. (2004), Brutsaert (1982), Winter et al. (1995), or Rosenberry et al. (2007)). The basic concepts of the three aforementioned groups and the mostly used formulas will be explained.

Dalton type equations (Aerodynamic approach)

The approaches based on Fick's first law (Eq. 2.20, 2.21) of mass transfer are called "Dalton type equations", as Dalton (1799) first expressed an equation for evaporation in the form of

$$Ev = (b_0 + b_1 \cdot v_a) \cdot (E_w - e_a), \tag{2.26}$$

where b_0 and b_1 are empirical constants,v_a is wind speed, E_w saturation vapour pressure of the water surface, and e_a vapour pressure of the air. The term in the first bracket is also called the wind function, f(v_a). Theoretical consideration suggested $b_0 = 0$, which was confirmed by several studies (Harbeck et al., 1952; Ficke, 1972; Brutsaert, 1982). To derive the constant b_1 various approaches were made. Either empirical approaches, or, like Brutsaert (1982), basic theoretical approaches are used. Brutsaert (1982) assumed a logarithmic wind profile and integrated between two heights, which results in

$$b_1 = K_E = \frac{0.622 \cdot \rho_a}{p \cdot \rho_w} \cdot \frac{\kappa^2}{(\ln(\frac{z_2 - z_d}{z_1 - z_d}))^2} \tag{2.27}$$

with air pressure, p, water density, ρ_w, the Kármán constant, κ=0.4, zero-plane displacement height, z_d, and the bulk transfer coefficient for evaporation, K_E. For the two heights he used the measurement height, z_m, as z_2, and the height at which the wind speed is zero as lower level, which is at $z_1 = z_d + z_0$. Brutsaert (1982) assumed for the roughness length of a water surface $z_0 = 2.3 \cdot 10^{-4}$ m. In this form the constant K_E does not consider atmospheric stability which supports or suppresses free convection and hence turbulent transport. The stability can be incorporated into the equation by the factors Φ such that

$$b_1 = K_{E,\Phi} = \frac{0.622 \cdot \rho_a}{p \cdot \rho_w \cdot \Phi_m \cdot \Phi_v} \cdot \frac{\kappa^2}{(\ln(\frac{z_m - z_d}{z_0}))^2}. \tag{2.28}$$

The factors Φ are related to stability by the Bulk-Richardson number

$$Ri_b = \frac{g \cdot (z_2 - z_1)(T_2 - T_1)}{0.5 \cdot (T_2 + T_1) \cdot (v_2 - v_1)^2}. \tag{2.29}$$

T_2 equals the air temperature at measurement height $T_a(z_m)$ and T_1 is the surface temperature, T_s. Negative Ri_b values represent stable conditions, positive Ri_b values represent unstable atmospheric conditions, and $Ri_b = 0$ neutral conditions. The Φ factors can be determined considering atmospheric stratification using the relations after Cline (1997):

$$\begin{array}{lll} Ri_b < -0.03 & \Phi_m = (1 - 18 \cdot Ri_b)^{-1/4} & \Phi_v = 1.3 \cdot (1 - 18 \cdot Ri_b)^{-1/4} \\ -0.03 \leq Ri_b \leq 0 & \Phi_m = (1 - 18 \cdot Ri_b)^{-1/4} & \Phi_v = (1 - 18 \cdot Ri_b)^{-1/4} \\ 0 \leq Ri_b \leq 0.19 & \Phi_m = (1 - 5.2 \cdot Ri_b)^{-1} & \Phi_v = (1 - 5.2 \cdot Ri_b)^{-1}. \end{array}$$

Other variations of the Dalton type equation suggest lake size as a critical parameter in K_E (Harbeck, 1962) or considered stability by $b_0 \neq 0$ like Rasmussen et al. (1995).

Energy balance approach

The energy balance approach is based on equation 2.18 and uses the Bowen ratio

$$\beta = \frac{H}{LE}. \tag{2.30}$$

H and LE are expressed by Eqs. 2.20 and 2.21, and it is assumed that D_H=D_{LE}.

Replacing specific humidity with the water vapour pressure (e_a) and moving from the differential form to differences leads to

$$\beta = \frac{c_p \cdot p}{0.622 \cdot L_v} \cdot \frac{T_s - T_a}{E_w - e_a} = \gamma \cdot \frac{T_s - T_a}{E_w - e_a}. \tag{2.31}$$

γ is called the psychometric constant. Using the so found relationship of H to LE yields the following equation for estimating evaporation

$$Ev = \frac{Rn - G - F_n - \Delta Q/\Delta t}{\rho_w \cdot L_v \cdot (1+\beta)}. \tag{2.32}$$

To calculate evaporation, net radiation, ground heat flux, net advected heat flux, heat storage, surface water temperature, air temperature, and saturation vapour pressure from the water and from the air have to be known. The advected heat flux can be neglected if the water volumes of the in- and outflow are small compared to the water body or if water temperatures are similar (Dingman, 2002). Also the ground heat flux can usually be neglected as the energy exchange at the bottom of the lake is small compared to the other terms of Eq. 2.32 (Henderson-Sellers, 1986). Some approaches choose Δt such that ΔQ is small and neglect the term as well. For most lakes this will be true for a time interval of 1 yr or when the lake becomes isothermal for an interval of 0.5 yr (Dingman, 2002).
If another time interval is chosen to calculate evaporation, the heat storage term has to be considered. The heat storage term describes the change in water temperature of the lake volume from one time step to another. To calculate it, water temperature and the lake volume at both time steps have to be measured. If these data are not available, a model for the heat storage term depending on the net radiation can be used (Duan and Bastiaanssen, 2015). Duan and Bastiaanssen (2015) proposed a model which considers the hysteresis of the heat storage term. The model has the form

$$\frac{\Delta Q}{\Delta t} = a \cdot Rn + b + c \cdot \frac{dRn}{dt}. \tag{2.33}$$

When using Eq. 2.33 in Eq. 2.32, evaporation for shorter timer intervals can be calculated as well.

Penman equation (Combination approach)

Penman (1948) was the first who combined the energy balance equation with the mass transfer (aerodynamic) equation to eliminate the need of surface water temperature data. He used Eq. 2.18, neglected the advected heat fluxes (F_n), the ground heat flux (G), and the heat storage term (ΔQ), and used the Bowen ratio (Eq. 2.31) and got as initial equation for the evaporation

$$Ev = \frac{LE}{\rho_w L_v} = \frac{Rn - \beta \cdot LE}{\rho_w \cdot L_v}. \tag{2.34}$$

With the assumption that the slope of the saturation vapour pressure versus temperature curve can be approximated as

$$\Delta = \frac{\mathrm{d}E}{\mathrm{d}T} \approx \frac{E_w - E_a}{T_s - T_a}, \tag{2.35}$$

the Bowen ratio can be rewritten to

$$\beta = \frac{\gamma\,(E_w - E_a)}{\Delta\,(E_w - e_a)}. \tag{2.36}$$

The relation remains true if e_a is subtracted from each term of the numerator on the right hand side

$$\beta = \frac{\gamma}{\Delta} \frac{[(E_w - e_a) - (E_a - e_a)]}{(E_w - e_a)}. \tag{2.37}$$

Penman (1948) then used Eq. 2.26 to replace $(E_w - e_a)$ and got

$$\beta = \frac{\gamma}{\Delta}\left(1 - \frac{(E_a - e_a) \cdot (b_0 + b_1 \cdot v_a)}{Ev}\right). \tag{2.38}$$

Substituting Eq. 2.38 into Eq. 2.34 and solving the equation for Ev yields

$$Ev = \frac{\Delta \cdot Rn + \gamma \cdot (b_0 + b_1 \cdot v_a) \cdot \rho_w \cdot L_v \cdot (E_a - e_a)}{\rho_w \cdot L_v \cdot (\Delta + \gamma)}. \quad (2.39)$$

Van Bavel (1966) replaced the empirical wind function $(b_0 + b_1 \cdot v_a)$ with Eq. 2.27 and got the following equation for evaporation

$$Ev = \frac{\Delta \cdot Rn + \gamma \cdot K_E \cdot v_a \cdot \rho_w \cdot L_v \cdot (E_a - e_a)}{\rho_w \cdot L_v \cdot (\Delta + \gamma)}. \quad (2.40)$$

The heat storage term can also be considered in this equations by replacing Rn with $(Rn - \Delta Q)$. Thus, the surface water temperature is no longer needed to calculate evaporation as long as the longwave outgoing radiation from the water surface is directly measured.

If it is not measured, longwave outgoing radiation can only be calculated using surface water temperature and the Stefan-Boltzmann equation (Eq. 2.15). Therefore, Kohler and Parmele (1967) suggested another approximation to eliminate the surface water temperature completely and derived the following parameters to calculate the net longwave radiation (L) and the psychometric constant (γ)

$$L' = L\downarrow - L\uparrow = \varepsilon_w \cdot L\downarrow - \varepsilon_w \cdot \sigma_B \cdot T_a^4 \quad (2.41)$$

$$\gamma' = \gamma + \frac{4 \cdot \varepsilon_w \cdot \sigma_B \cdot T_a^3}{K_E \cdot \rho_w \cdot L_v \cdot v_a}. \quad (2.42)$$

Priestley-Taylor equation (Combination approach)

To further reduce the amount of necessary measurements to calculate evaporation from a lake surface, Priestley (1972) replaced the second term of the Penman equation with a dimensionless empirical multiplier, the Priestley-Taylor coefficient $c_{PT} = 1.26$, and got

$$Ev = c_{PT} \frac{\Delta \cdot Rn}{\rho_w \cdot L_v \cdot (\Delta + \gamma)}. \quad (2.43)$$

Selection of an Equation

The choice of an equation to estimate evaporation often depends on the availability of meteorological measurements for an investigation site and on the extent of the data set, or on the possibility to conduct own measurements. Table 2.1 gives an overview of the variables necessary to solve the equations when all terms are considered. The Dalton type equation needs only four input parameters, which is much less compared to the equations considering the energy balance, but the main disadvantage of the Dalton type equation is the need of an empirical wind function, which has to be known for the investigation site. Otherwise, the more general approach after Brutsaert (1982) has to be used. The equations considering the energy balance need generally much more variables, but as already mentioned some of the terms of the energy balance can be neglected under certain circumstances. For larger lakes the ground heat flux (G) and the net advected energy flux (F_n) can often be neglected (Dingman, 2002) and the heat storage term can be neglected if the averaging period is long enough, e.g. 1 yr. This reduces the amount of input variables to four for the energy balance and Penman approach and to three for the Priestley-Taylor equation.

When conducting own measurements, not only the amount of necessary variables, but also the complexity of the measurement setup for a specific variable has to be considered. Wind speed and air temperature can often be measured with enough precision at the shore line, but the surface water temperature and the longwave outgoing radiation require a measurement setup directly above the water. Finally, the time steps for which evaporation shall be calculated have to be considered. For short averaging intervals the Dalton type equation is generally the best choice, as it calculates instantaneous evaporation amounts, and the error on the evaporation estimates grows larger with larger averaging intervals (Dingman, 2002). For the equations considering the energy balance time intervals of at least 7 d are recommended in literature (e.g. Dingman (2002)), because the energy balance closure is often not fulfilled for shorter averaging intervals. For choosing an equation to estimate evaporation, all these discussed factors have to be taken into account. If all variables can be measured, the Penman equation is a good choice as it considers both the energy an the mass transfer. If one of the energy components can not be measured or if short time intervals are investigated, the aerodynamic approach is most likely the better choice, even though the empirical wind function bears a strong uncertainty.

Table 2.1: Overview of the necessary variables to calculate evaporation.

	v_a	$f(v_a)$	T_s	T_a	R_n	G	F_n	$\Delta Q/\Delta t$
Dalton type equation	X	X	X	X	-	-	-	-
Energy balance approach	-	-	X	X	X	(X)*	(X)*	X
Penman equation	X	X	-	X	X	(X)*	(X)*	X
Priestley-Taylor equation	-	-	X	X	X	(X)*	(X)*	X

*Terms, which are usually not considered in calculations.

3 Research Area and Data Base

In order to study the coupled atmospheric, hydrological, and lithospheric processes connected to environmental risks, water availability, and climate change in the Dead Sea region, the Helmholtz Virtual Institute DESERVE was founded (Kottmeier et al., 2016). DESERVE is designed as a cross-disciplinary international project which focuses on the understanding of interconnected environmental processes, e.g. the understanding of all water budget components and the sinkhole formation. To achieve these goals, a cross-disciplinary monitoring network was established, and several research activities investigating the different processes were conducted. The research questions concerning the specific atmospheric conditions prevailing in the Dead Sea valley and their influence on the land surface - atmosphere exchange processes are addressed in this thesis.

3.1 Characteristics of the Research Area

Geography

The Dead Sea basin is part of the Jordan Rift Valley, which is a north-south oriented depression zone extending from the northern Israeli border to the Gulf of Aqaba in the south. The Dead Sea basin is the lowest area of this depression zone. It is approximately 150 km long and 15 to 17 km wide in the centre (Fig. 3.1). It is confined by the Judean Mountains to the west, which have a mean mountain ridge height of 895 m AMSL in the northern part (31.3°-31.8° N) and 615 m AMSL further to the south, and by the Moab Mountains to the east, with a mean mountain ridge height of 950 m AMSL in the north and 1150 m AMSL in the south.

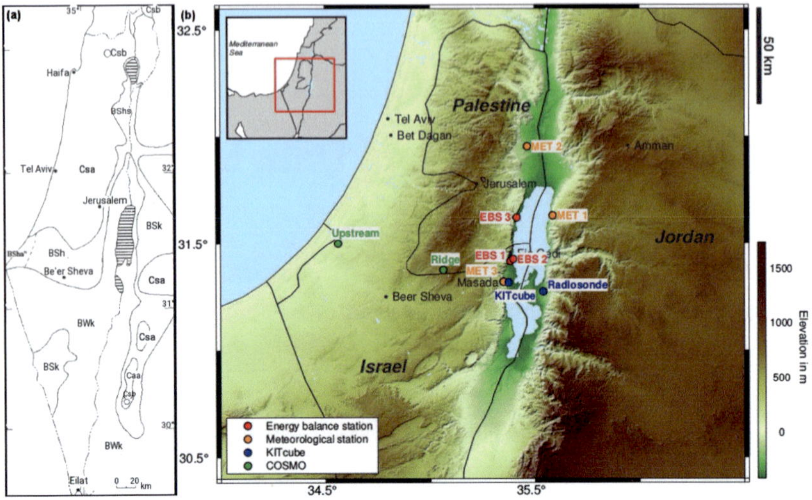

Figure 3.1: Map of Koeppen climate classification of Israel. Cs - Mediterranean climate; Bs - semi-arid climate; BW - arid climate; a - hot summer; b - warm summer; h - hot arid; k - cold arid. Modified after Goldreich (2003) (a). Topographic map of the research area with the measurement sites (b).

The Dead Sea is a hypersaline terminal lake with currently $367\,\mathrm{g\,l^{-1}}$ total dissolved solids (Siebert et al., 2014). Since the 1950s, the lake level dropped over 30 m, from −395 m AMSL to the current −429.9 m AMSL (Givati and Tal, 2016). This lake level decline is mainly caused by anthropogenic interferences. Especially the over-exploitation of the water resources of the Jordan River and its watershed, which was the main inflow to the Dead Sea, led to a reduction of the inflow by 90 %. The severe lake level decline caused a split of the Dead Sea into a northern basin, which has a surface area of approximately $600\,\mathrm{km}^2$ and a depth of up to 300 m (Hall, 1997), and a southern basin. The southern basin would have dried up in 1976 (Steinhorn et al., 1979), but as it is used as industrial evaporation ponds, the companies pump water from the northern basin to the southern ponds to proceed with their production of minerals and potash (Gertman and Hecht, 2002). The ponds have a surface area of $280\,\mathrm{km}^2$

and are only about 10 m deep. The area between the lake and the eastern and western mountain chains is rocky desert and covers roughly 580 km^2. When freshwater springs emerge along the shore line, sufficient water is available for plants to grow. The total area of these vegetated areas is very small compared to the area covered by water or desert, but these vegetated areas are very important for the diversity of the local ecosystems.

Weather and Climate

The region can be divided into three climate zones according to the Koeppen-Geiger classification (Fig. 3.1 a). A Mediterranean climate prevails over the coastal plains from the Galilee in the north to the northern border of the Negev in the south. The Negev and the Dead Sea valley are hot desert and in between those two climate zones a semi-desert region is located (Goldreich, 2003). As a result of this strong climatic gradient the whole region is highly sensitive to climate change (Alpert et al., 1997; Smiatek et al., 2011). On an annual scale the climatic conditions in Israel can be subdivided into a dry period in summer with no precipitation (June-Aug), a rainy season in winter with considerable rainfall, higher mean relative humidity and lower maximum mean temperature, and the transition seasons in spring and autumn, which have some rainfall, lower relative humidity and highest monthly maximum mean temperatures (Jaffe, 1977). The climate graph in Fig. 3.2 illustrates the mean climatic conditions and also shows the differences between the climate zones. In Tel Aviv, where Mediterranean climate prevails, precipitation amounts are considerably higher than in the semi-desert regions around Beer Sheva or in the desert near Eilat. Annual mean precipitation amounts of 580 mm in Tel Aviv compared to only 195 mm in Beer Sheva and only 32 mm in Eilat show the strong climatic gradients within the country. Larger temperature differences can mainly be seen between Eilat and the other two cities. Eilat is located furthest to the south and shows the highest mean monthly temperatures in all months, with a minimum of 15 °C in January and over 30 °C in summer.

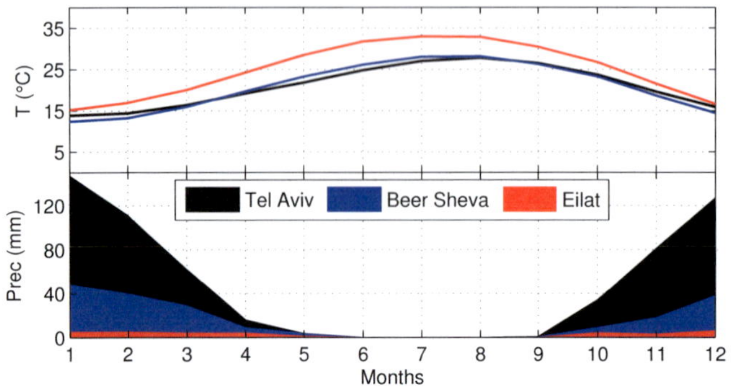

Figure 3.2: Climatic conditions of Jerusalem, Beer Sheva, and Eilat at the Red Sea. Monthly mean air temperature and precipitation for the climate normal period from 1981 to 2010.

Tel Aviv and Beer Sheva have similar mean monthly temperatures, but Tel Aviv has higher temperatures in winter and lower ones in spring compared to Beer Sheva. This is caused by the tempering effect of the Mediterranean Sea leading to a reduced temperature amplitude in Tel Aviv with only 13.9 °C compared to 15.7 °C in Beer Sheva and 17.7 °C in Eilat. In the Dead Sea valley itself average annual air temperature is 25.9 °C (Hecht and Gertman, 2003) and maximum daily air temperature frequently exceeds 40°C in summer. Annual precipitation amounts are about 70 mm (Goldreich, 2003), with no rain in summer and only a few precipitation events in winter.

The division in dry, wet and transition seasons is coupled to the large scale synoptic forcing. In the transition seasons and in winter the Icelandic low at about 60° N latitude extends southward and in the Mediterranean region a frontal zone may form between the cool and moist air masses coming from Europe and the warm and dry air masses from the Saharan desert (Goldreich, 2003). Cyclogenesis in this frontal region over the Mediterranean, especially south of Cyprus, causes 90% of the winter rainfall in Israel (Goldreich, 2003). Another synoptic system, which influences Israel in the transition seasons and in winter, is the Red Sea Trough (RST) (Fig. 3.3 a). The RST forms as a northward

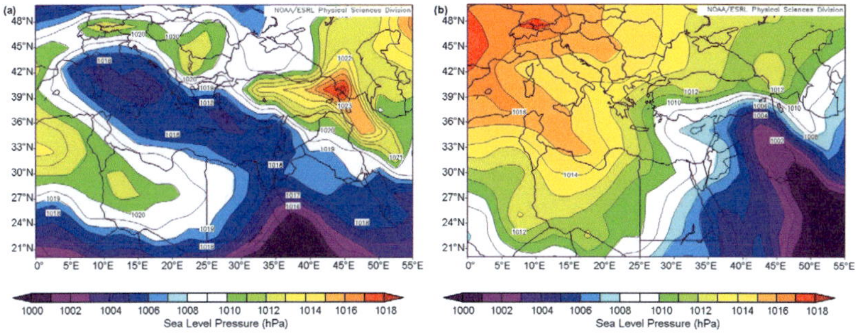

Figure 3.3: Mean sea level pressure for winter (a) and summer (b) for the climate normal period from 1981 to 2010. Images provided by Physical Sciences Division, Earth System Research Laboratory, NOAA, Boulder, Colorado, from http://www.esrl.noaa.gov/psd/.

movement of the Sudan low, which belongs to the inner-tropical convergence zone. Depending on its exact location, the RST leads to a westerly or north-westerly flow over Israel when its axis is located east of Israel and to an easterly or south-easterly flow when its axis is located west of Israel. In summer, Israel is located on the edge of the Persian Trough. This warm trough is an extension of the monsoon low and extends from the Persian Gulf, to the north-west over Iraq, to the north-eastern Mediterranean Sea where it meets the Mediterranean ridge (Fig. 3.3 b) (Bitan and Sa'Aroni, 1992; Alpert et al., 1990). The Persian Trough appears only in the lower boundary layer up to 850 hPa. It results in a persistent north-westerly flow in the lower levels towards Israel, also called the Etesian flow. Above, the Mediterranean ridge causes upper air subsidence leading to a stratification of the air mass with a strong inversion above the ABL. This leads to hot and dry weather conditions.

3.2 Measurement Sites and Instrument Specifications

Within DESERVE, an extensive field campaign to investigate the atmospheric conditions was conducted. The concept of the field campaign comprises an atmospheric long-term monitoring network and two embedded SOPs, which

were conducted in August and November 2014. The long-term monitoring network, consisting of three Meteorological Towers (METs) and three sophisticated Energy Balance Stations (EBSs), was especially designed to provide high resolution and quality data of the energy balance components of the land and water surface as well as of the near-surface atmospheric conditions in and around the valley over a long period of time. Data from March 2014 until March 2015 were used to study diurnal but also intra-annual variations of the energy balance components as well as the frequency of occurrence of local wind systems. With this monitoring network the near-surface conditions are covered sufficiently, but it can not provide information about the three-dimensional structure of the wind systems, their driving forces or their impact on the valley atmosphere. Therefore, vertical profiles of humidity, temperature, horizontal wind speed and direction, as well as integrated water vapour and aerosol measurements, are necessary. These data sets were measured during the SOPs with various instruments, e.g. lidar systems, radiosondes and GPS. For information about the atmospheric conditions outside of the valley, data from the operational measurement network of the Israel Meteorological Service (IMS) were used.
For the calculation of the latent heat flux and the saturation vapour pressure, special characteristics of the saline water, such as the latent heat of vaporisation and the activity of water, are necessary. Those parameters were measured in the laboratory of the Institute of Physical and Theoretical Chemistry at the University of Wuppertal in Germany.

3.2.1 Long-term Monitoring Network

Measurement Locations

The three METs are installed around the Dead Sea (3.1 b). The first tower (MET1) is located at the Jordanian side of the Dead Sea at 124 m AMSL and was installed in February 2014. The second tower (MET2) is located in al-Auja in Palestine at −228 m AMSL and was installed in November 2014, and the third tower (MET3) is located at the western side of Masada at −7 m AMSL

and already provides data since 2007. The three fully equipped EBSs were installed in February 2014 and are located at three characteristic sites at the western shore of the Dead Sea to capture the energy balance components of the different surface types in the valley, water, bare soil, and vegetation, under similar atmospheric conditions. EBS1 measures the conditions over bare soil, EBS2 measures the conditions over the water and EBS3 measures the conditions over a vegetated area.
EBS1 is located approximately 3 km south of Ein Gedi, at −393 m AMSL, and 1 km west of the shore line (Fig. 3.1). The land surface consists of gravel and sand (Fig. 3.4 a). The second measurement site (EBS2) is located 1.6 km to the north-east of EBS1, on the tip of a headland at the western shore of the Dead Sea, at −428 m AMSL (Fig. 3.1, 3.4 b). The headland is 214 m long and is surrounded by water from 300° to 260° (insert in Fig. 3.4 b). This means that not all measured turbulent fluxes represent the fluxes from the water surface, but the installation of the station at the shoreline has several advantages compared to an installation on a floating pontoon in the middle of the lake. First, the station is build on solid ground and the instruments do not move. If the station would be installed on a floating pontoon, an extensive correction of the wind data would be necessary to remove the movement of the pontoon from the data. Especially in high wind situations with an increased wave activity, the data would be heavily affected. The second advantage of the location at the shoreline is the longer maintenance interval. On a pontoon a maintenance interval of 7 days or less would be necessary, as because of the sea spray of the highly saline water, the instruments would be covered with a salt layer after a couple of days. Especially the radiation and water vapour measurements would be affected by this salt layer. Additionally to the technical difficulties the tense political situation has to be considered. The Dead Sea area is under the administration of Israel, the Palestinian Authorities, and Jordan, which often makes it difficult to receive permission for measurements in the area. Thus, the installation on land was favoured over the installation on a pontoon to assure good data quality and realise the measurements. The third station, EBS3, is located further to the

Figure 3.4: Overview of the measurement sites. The measurement site of EBS1 is over bare soil (a), EBS2 is located at the shoreline (b) on a headland with an opening angle of 40° (insert in b), and EBS3 is located within vegetation (c).

north in an area mainly covered by reed (Fig. 3.1). It is mounted in a clearing surrounded by reed which is up to 5 m high (Fig. 3.4 c).

Instrumentation

At all stations, standard meteorological variables, such as 2 m temperature and humidity, air pressure, shortwave and longwave radiation components of the upper and lower half space, are measured and averaged over 10 min intervals. Precipitation is measured and summed over the same time period. Additionally, soil temperature and soil moisture are measured and also averaged over 10 min at the three METs as well as at EBS1 and EBS3. Supplementary measurements of air temperature in 6 m height were performed at the EBSs only. With a temporal resolution of 20 Hz, all three wind components are measured at the METs with a Gill sonic anemometer at 10 m height, and at the three EBSs all three wind components, sonic temperature, and water vapour are measured at 6 m height with an Integrated Gas Analyser and Sonic Anemometer (IRGASON).

Figure 3.5: Experimental setup of the radiation measurements over the water at the beginning of the experiment (a) and at the end of the experiment (b and c).

Radiation Measurements over the Dead Sea Water Surface

To be able to calculate the reflected shortwave radiation from the water surface, additional measurements of the radiation fluxes of the water surface were performed in a short-term experiment. For a period of 8 days, from 1 until 8 November 2014, the shortwave upward and downward radiation flux was measured 1 m above the water surface (Fig. 3.5). At the end of the experiment the instrument was already covered with a salt layer caused by the sea spray (Fig. 3.5 b and c).

Integrated Water Vapour by GPS

At the energy balance stations EBS1 and EBS2, Global Positioning System (GPS) receivers were installed by the German Research Centre of Geoscience (GFZ) in June 2014. It was already shown in several studies that it is possible to use the GPS network to monitor atmospheric water vapour (Bevis et al., 1992; Rocken et al., 1997). The measurements are based on the fact that the GPS signal is delayed by the ionosphere and the troposphere on its way to the GPS receiver at the ground. The ionospheric refraction is dispersive and can be removed by a linear combination of two frequencies (Dick et al., 2001). The tropospheric delay depends on pressure, temperature and humidity. Especially water vapour strongly influences the delay of the GPS signal and can therefore be used to retrieve the amount of integrated water vapour (IWV) in the atmosphere

(Rocken et al., 1995). The Zenith Total Delay (ZTD) can be partitioned into two components: the Zenith Hydrostatic Delay (ZHD), caused by the delay in dry air, and the Zenith Wet Delay (ZWD), caused by the water vapour

$$\mathrm{ZTD} = \mathrm{ZHD} + \mathrm{ZWD} \tag{3.1}$$

The signal delay also depends on the elevation angle and is smallest in the zenith direction (Dick et al., 2001). This dependence can be described by mapping functions $\mathrm{m_{hyd}}$ for the hydrostatic delay and $\mathrm{m_{wet}}$ for the wet delay. The tropospheric delay ($\Delta trop$) can be described by

$$\Delta trop = \mathrm{m_{hyd}ZHD} + \mathrm{m_{wet}ZWD}. \tag{3.2}$$

The ZWD can then be transformed into integrated water vapour estimates with a very high accuracy, using surface measurements of pressure and temperature (Rocken et al., 1995),

$$IWV = \Pi(T_m)\mathrm{ZWD}, \tag{3.3}$$

where $\Pi(T_m)$ is a conversion factor after Askne and Nordius (1987), which depends on the mean temperature of the atmosphere, T_m, above the sensor. This factor is about 1/6, but due to the uncertainties in determining T_m, it can vary up to 20 % (Bevis et al., 1994). The conversion of the GPS data into IWV was done by Galina Dick (GFZ Potsdam) using surface data from the energy balance stations with a temporal resolution of 15 min.

3.2.2 Special Observation Periods

For the SOPs the instruments of the mobile observation platform KITcube were installed near Masada, at −350 m AMSL (Fig. 3.1). The site is located 1.4 km east of Masada mountain, which is 59 m AMSL high and 2.7 km west of the Dead Sea shoreline (Fig.3.6). The southern ponds are 9 km south-east of the KITcube location.

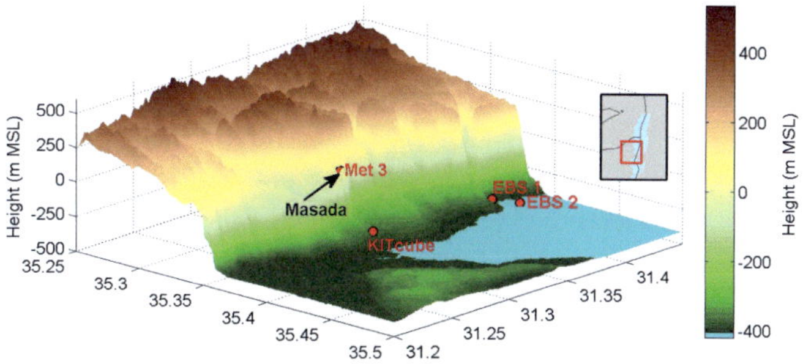

Figure 3.6: Topographic map of the southern part of the Dead Sea with the location of the SOP measurement site.

The instruments were installed on a terrain consisting of bare soil with small dry canyons. Additionally, a radiosonde station was located on the eastern side of the Dead Sea at Al Mazraa, at −250 m AMSL (Fig. 3.1). KITcube consists of several in-situ and remote sensing instruments which provide information about the energy exchange at the Earth's surface, turbulence, and mean conditions in the ABL and whole troposphere respectively, and about cloud and precipitation properties. A complete description of KITcube is given by Kalthoff et al. (2013); here, only the instruments which are used in this thesis are described.

Energy Balance Station

Standard meteorological variables, such as 2 m temperature, humidity, wind speed, wind direction, air pressure, precipitation, radiation temperature of the surfaces, and shortwave and longwave radiation components of the upper and lower half space, are measured with an integration time of 10 min. Additionally, water vapour and the three wind components are measured with a fast infrared hygrometer and an ultrasonic anemometer with a temporal resolution of 20 Hz.

GPS

A temporary GPS sensor was installed at the KITcube location, and the IWV was calculated using surface measurements from the energy balance station. The conversion of the GPS data was performed by Galina Dick (GFZ Potsdam) using algorithms described by Dick et al. (2001) and Gendt et al. (2004). The temporal resolution was 15 min.

Wind Lidars

Light detection and ranging (lidar) systems are active remote sensing instruments to measure atmospheric properties at some distance from the instrument. The principle of active remote sensing and the specifics of lidar systems are described below following the description of Banta et al. (2013). In active remote sensing the instruments emit electromagnetic radiation and detect the energy scattered back at 180° from targets in the atmosphere as a function of time delay. The scattering targets are determined by the wavelength. Lidar systems use a narrow laser beam with wavelengths between 200 nm and 10 µm, such that molecules and aerosols are scattering targets of the lidar. The strength of the received signal depends on the strength of the emitted signal, attenuation losses, the backscatter cross section, and the distance of the scattering targets from the instrument. There are two reasons for attenuation loss. First, absorption by molecules or particles and second, scattering of the energy in directions different than 180°, meaning scattering out of the beam instead of backwards towards the receiver. Liquid water, such as in fog or liquid water clouds, attenuates the energy at lidar wave lengths in such a way that the penetration depth of the lidar signal is very limited. The backscatter cross section is a defined cross-sectional area which represents the actual backscattered energy by all targets in a pulse volume. Except for the backscattered signal, the instrument measures background noise, which often dominates the return signal. The Signal-to-Noise Ratio (SNR)

$$SNR = 10 \log_{10} \frac{\sigma_s}{\sigma_N}. \tag{3.4}$$

is the ratio of the standard deviation of the received signal strength, σ_s, to the standard deviation of the noise, σ_N, and is given in decibels (dB). As noise is a random process, averaging the received signal over several pulses increases the detectability of the signal, i.e. the SNR is increased.

Wind lidars, so-called coherent Doppler lidars, measure radial velocity along the beam by calculating the frequency difference (Δf) between emitted and received signal, which is caused by the movement of the target, i.e. the Doppler shift. Therefore, the time series of the received energy is transformed into the frequency domain by a Fourier transform, and the radial velocity (v_r) can be calculated with

$$v_r = \frac{1}{2}\frac{c \cdot \Delta f}{f}, \tag{3.5}$$

with the phase speed, c, and frequency, f. The calculated radial velocity is an average over the pulse volume and can not resolve variations within the volume. To obtain horizontal wind speed, an Velocity-Azimuth-Display (VAD) algorithm can be applied to the data.

One wind lidar used during the SOPs was a two-axis scanning pulsed 2 µm wind lidar from Lockheed Martin called 'WindTracer', with a peak power of 4.5 kW and a pulse repetition frequency of 500 Hz (Träumner et al., 2012). The effective pulse length and hence the minimum spatial resolution is 56 m, and the beam width at the laser exit is 75 mm. A more detailed description of the instrument can be found in Träumner (2012). For comprehensive measurements of the wind field, Range-Height-Indicator (RHI) scans were performed for an elevation ranging from 0 to 180° at azimuth angles of 15, 62, 119, and 172°, and Plan-Position-Indicator (PPI) scans of 360° azimuth were performed for elevation angles of 0.2, 5, 15, and 75°. Only for the lowest PPI at 0.2° the azimuth was restricted to about 180° because of the mountains to the west. The maximum rotation speed was 20° s^{-1} and the minimum detection range of the instrument was 400 m AGL. The duration of a PPI scan was about 1 min and of a RHI 30 s. The performance of the instrument was good and a SNR of -7 dB was used as a quality criterion.

To cover the gap between the surface and the minimum detection range, a second wind lidar from Leosphere, called 'Windcube', with a wavelength of 1.54 µm, was operated at the KITcube location. The detection range was from 40 to 600 m AGL and the spatial resolution was 20 m. The lidar worked in a 4-point stop and stare mode at a fixed elevation angle of 75.2° with and integration time of 7 s, and the horizontal wind profile was then calculated using a VAD algorithm after Browning and Wexler (1968). A carrier-to-noise ratio of -23 dB was used to ensure data quality.

Radiosonde

To obtain high resolution vertical profiles of temperature, humidity, wind speed, and wind direction, two radiosonde systems from Graw were operated during Intensive Observation Periods (IOPs). The radiosondes were launched in an 2 h interval, always from Friday 13 Local Time (LT = Coordinated Universal Time (UTC)+2 h) until Sunday 6:00 LT. The radiosondes could not be launched on other days due to restrictions from the air space control. The temporal resolution of the raw data is 1 s. Between the ground and 3 km AGL data are averaged over 2 s, between 3 km and 10 km AGL data are averaged over 4 s, and above 10 km AGL data are averaged over 8 s. Operational radiosondes were launched at Bet Dagan every day at 2:00 LT (0:00 UTC) and 14:00 LT (12:00 UTC) by the IMS.

3.3 Laboratory Measurements

For the calculation of the latent heat flux of the Dead Sea water surface the latent heat of vaporisation of the water is required (Sec. 2.3). The latent heat of vaporisation of freshwater depends only on water temperature, and an equation is already given in Sec. 2.3. For saline water like the Dead Sea brine, the chemical composition and the salinity has to be considered additionally. Therefore, the temperature dependence of the latent heat of vaporisation and the activity of water, which is needed for the calculation of the saturation vapour pressure,

Figure 3.7: Experimental setup to measure the latent heat of vaporisation of Dead Sea water.

were measured for the highly saline water of the Dead Sea. The measurements are based on the principle that two phases of a substance can only coexist at equilibrium at a certain pressure, when the temperature is held constant. This pressure, called saturation vapour pressure, can be measured as a function of temperature. The energy needed for the phase transition under conditions of constant temperature and pressure is called molar latent heat of vaporisation and can be calculated indirectly from the temperature and pressure dependence using the Clausius-Clapeyron equation.

The experiment was performed at the Institute of Physical and Theoretical Chemistry at the University of Wuppertal with the help of Prof. Dr. Jörg Kleffmann. A water probe which was taken at the measurement site of EBS2 at the end of 2014 was used to conduct the experiment. The water probe was filled in a u-tube which was connected to a capacitance manometer (Baratron; manufactured by MKS) on the one side and to a pump on the other side (Fig. 3.7). The Baratron was calibrated against a McLeod manometer with a precision of 0.01 mbar. The u-tube was put in a digital controlled, refrigerated circulation bath with a temperature stability of 0.04 °C. A calibrated thermometer was additionally used to measure the water temperature. The u-tube was evacuated with the pump and then measurements were undertaken for a temperature range of -3 to 33 °C. The temperature was increased stepwise, and the saturation vapour pressure was always measured after the equilibrium state was reached.

4 Methods of Data Analysis

In Chap. 5 an investigation of the wind systems and their characteristics will be presented. To obtain an objective analysis of the layer characteristics, an automatic detection method is necessary. The work flow of the used algorithm is described in Sec. 4.1. For the analysis of the energy balance components in Chap. 6 the eddy covariance measurements of the turbulent fluxes need an extensive post-processing. Firstly, measured values have to be checked on plausibility. Secondly, several corrections have to be applied. A correction to account for a possible influence of the horizontal wind components on the vertical wind component, when the instrument is not perfectly levelled. Corrections for the influence of humidity on the temperature measurements, for spectral losses, and for the temperature and humidity dependency of the water vapour measurements. Furthermore, the data have to be tested on stationarity of the fluctuations within an averaging interval and on turbulent conditions, because these are prerequisites for the eddy covariance method. The steps of the post-processing are described in detail in Sec. 4.2. To account for the special composition of the Dead Sea water, i.e. extremely high salinity, several assumptions, which are often applied for less saline water, such as for the oceans, are not valid. The use of the latent heat of vaporisation for freshwater is not possible, and the reduced activity of the water has to be considered when calculating the saturation vapour pressure. Additionally, methods have to be found to account for variables which were not directly measured, i.e. surface water temperature and outgoing radiation from the water surface. Therefore, in Sec. 4.3 measurement results of the required parameters and methods to provide the not directly measured variables are presented. In Sec. 4.4 the general regression approach and the method to validate the regression model are described, which are used in Sec. 6.2 to predict the turbulent fluxes from the water surface.

4.1 Detection of Wind System Characteristics from Lidar Scans

In Chap. 5 the three-dimensional structure of thermally and dynamically driven wind systems will be analysed. These wind systems determine the properties of the atmospheric layer in which they occur, e.g. temperature, humidity, aerosol content, wind speed, or wind direction. Each atmospheric layer has its characteristic properties and a sharp gradient of these quantities often exist between the layers. Therefore, these gradients can be used to identify the vertical extent of the wind systems investigated, and further properties of the wind systems, such as their duration, mean wind speed, mean layer height, and mean maximum wind speed, can be calculated. RHI scans from the two-axis scanning lidar along the main wind direction of the wind systems were used to calculate the aforementioned properties of the wind systems. From each RHI scan the horizontal component of the radial velocity was calculated and data in the range from 2 to 3 km away from the lidar were averaged to gain a vertical profile of the horizontal radial wind component with a vertical resolution of 20 m. These profiles were used for time-height cross-sections of the particular wind systems and the layer characteristics were calculated as follows.

For the along-valley flows, the layer height was determined as the height where the horizontal radial wind component changed the sign. This equals the top of the near surface branch of the circulation, or the bottom of the reverse flow aloft, respectively. If the reverse flow was not distinguishable from the large-scale flow, the layer height was determined as the height where the wind speed first started to increase again after an elevated local wind speed minimum was detected. DWFs do not have a closed circulation, but they normally contain a different air mass which has a different temperature, humidity, and aerosol content than the air in the valley.

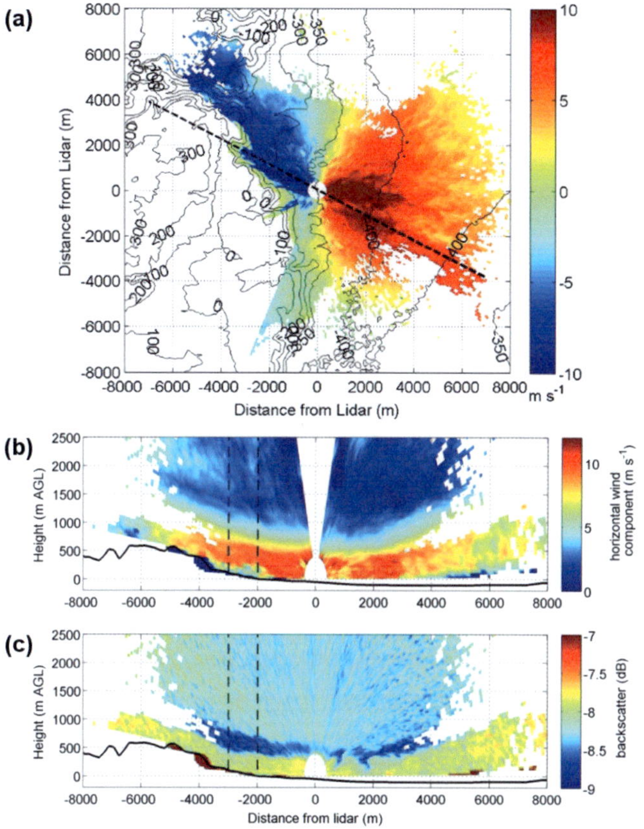

Figure 4.1: PPI scan of the radial velocity at 5° elevation at 16 August 2014, 19:01 LT (a). Blue colours indicate flow towards the lidar and red colours flow away from the lidar. The dashed line indicates the direction of the RHI scans of the horizontal component of the radial velocity shown in (b) and of the backscatter signal shown in (c) (note the different colour scales). RHI scans were performed at 16 August 2014, 19:06 LT.

As the lidar backscatter is sensitive to the aerosol content, a smoothed backscatter profile can be used additionally to the wind speed profile to determine the layer height. In a first step the height of the strongest wind speed gradient above the near surface wind speed maximum was identified.

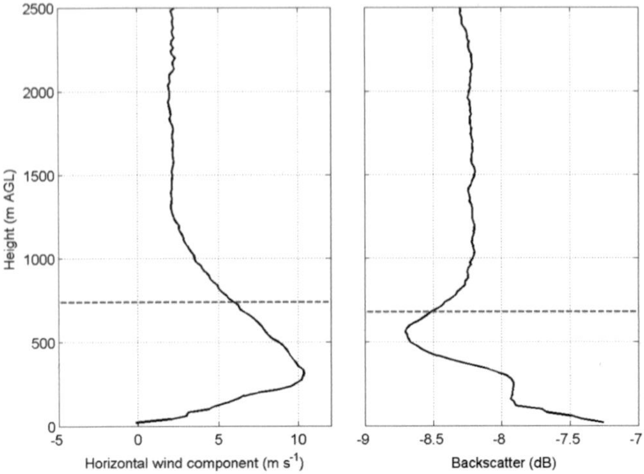

Figure 4.2: Vertical profiles of the horizontal radial wind component and of the backscatter signal averaged between the black dashed lines in Fig. 4.1 (b) and (c). RHI scans were performed at 16 August 2014, 19:06 LT.

Then the height of the strongest backscatter gradient was extracted and the heights were compared. If the height difference between the two derived heights was more than 100 m, a comparison to the layer height of the preceding time step was performed, and the closest match was used to determine the layer height.

An example of a strong DWF event on 16 August 2014 is shown in Fig. 4.1 to illustrate the detection algorithm. The wind field shows a strong northwesterly flow as can bee seen in the PPI scan of the radial velocity (Fig. 4.1 a). The RHI scan of the calculated horizontal component of the radial velocity along the dashed line (Fig. 4.1 a) shows very high wind velocities in a layer between the surface and approximately 1000 m AGL (Fig. 4.1 b). Above is a layer with very low velocities. Those two layers can also be identified in the corresponding backscatter RHI scan (Fig. 4.1 c). The DWF is characterised by a strongly changing backscatter signal with high values near the surface and a minimum in the upper part of the DWF. The layer above has a relatively uniform backscatter. The profiles of the horizontal radial wind component and the backscatter signal averaged over a horizontal distance of 1 km between the two black dashed lies in Fig. 4.1 (b) and (c) are shown in Fig. 4.2. The wind profile has a jet-like structure with a maximum around 340 m AGL, and a uniform layer starts at around 1200 m AGL. The backscatter profile shows a strong decrease until it reaches its minimum slightly above the wind speed maximum at around 580 m AGL. The heights of the layer detected using the backscatter and wind speed gradients are 680 m and 740 m, respectively. In this case the detected heights are slightly below the heights where the profiles are becoming constant again, but from a manual inspection of several cases the use of the strongest wind and backscatter gradients yields the best results for an objective layer height detection and a reasonable calculation of the layer characteristics.

4.2 Post-processing of Eddy Covariance Data

Post-processing of eddy covariance data is important, as the instrumentation and the measurements generally do not fulfil completely the theoretical concept of the eddy covariance (Sec. 2.3.1) and its assumptions, e.g. stationary conditions over the averaging period or horizontal homogeneity. Only after extensive post-processing good data quality and small flux errors can be assumed. The post-processing was done with the software package TK3 (Mauder and Foken, 2011), which allows the individual selection of the corrections. The following post-processing steps were applied to the data set used in the analysis.

4.2.1 Plausibility Tests

Data were first checked for plausibility and were rejected when they crossed the thresholds shown in Table 4.1. Secondly, a spike detection algorithm after Mauder et al. (2013) was applied. This algorithm is based on the Median

Absolute Deviation (MAD) of a quantity, and a data point x_i is detected as a spike when it is not in the range of

$$\langle x \rangle - \frac{c_t \cdot MAD}{0.6745} \leq x_i \leq \langle x \rangle + \frac{c_t \cdot MAD}{0.6745}. \tag{4.1}$$

$\langle x \rangle$ is the median of x, $MAD = \langle |x_i - \langle x \rangle| \rangle$, and c_t is a threshold which is set to 7, according to Mauder et al. (2013).

Table 4.1: Thresholds for plausibility test used for the post-processing of the flux data.

Quantity	Lower limit	Upper limit	Unit
u-wind component	-50	50	$\mathrm{m\,s^{-1}}$
v-wind component	-50	50	$\mathrm{m\,s^{-1}}$
w-wind component	-10	10	$\mathrm{m\,s^{-1}}$
sonic temperature	-20	100	°C
absolute humidity	0	35	$\mathrm{g\,m^{-3}}$

Gaps in the time series due to the plausibility check were treated as missing values and no interpolation of the data to close the gaps was applied. If after the plausibility test more than 10 % of the data in a time interval of 30 min were missing, no flux calculations were performed for this time interval, and the sensible and latent heat fluxes were set to missing values for this time step.

4.2.2 Corrections

Planar Fit Correction

Generally, the sonic anemometer can not be levelled perfectly that the vertical axis is perpendicular to the surface. Thus, vertical wind measurements can be affected by the other wind components, resulting in a non-zero mean vertical wind (Rebmann et al., 2012). Therefore, the planar fit method after Wilczak et al. (2001) was applied. Wilczak et al. (2001) proposed the following procedure. First, a mean stream line plane is calculated from the whole data set. As a result of rotating the coordinate system around the y-axis, the z-axis is positioned perpendicular to this plane, and the mean vertical wind is 0. Another rotation around the z-axis aligns the coordinate system with the mean wind direction which results in $\overline{v} = 0$.

Temperature Correction

Sonic anemometers measure the speed of sound, which depends on air temperature and humidity. To convert the buoyancy flux ($\overline{w'T'_{sonic}}$) to sensible heat flux, ($\overline{w'T'_a}$), Schotanus et al. (1983) proposed the following correction

$$\overline{w'T'_a} = \overline{w'T'_{sonic}} - 0.51 \cdot \overline{T_a} \cdot \overline{w'q'}, \tag{4.2}$$

which is also known as the SND-correction after the authors Schotanus, Nieuwstadt, and De Bruin (1983). Fig. 4.3 (a) shows an exemplary diurnal distribution of the differences of the buoyancy and the sensible heat flux at the three energy balance stations. The differences range from -5 up to 25 W m^{-2}. At EBS1, where the sensible heat flux was high with a maximum of 335 W m^{-2} (Fig. 4.3 c), the corrections play a minor role, but at EBS2, where the maximum sensible heat flux was 35 W m^{-2} on that day, the correction term changes the results considerably, as the correction term is in the order of 23 % of the measured buoyancy flux. At EBS3 the sensible heat flux is comparable to the heat flux at EBS1, but the correction term is higher and is already in the range of 10 % of the sensible heat flux. Thus, the SND correction is especially important when the sensible heat flux is already small or when the humidity fluctuations are very large.

WPL-Correction

Most of the gas analysers, including the IRGASON, measure the amount of water vapour in the air as a molar density and not as a mass mixing ratio. Therefore, the water vapour data are influenced by temperature and humidity changes. The WPL-Correction, named after Webb, Pearman, and Leuning (1980), applies corrections for such fluctuations of humidity and temperature.

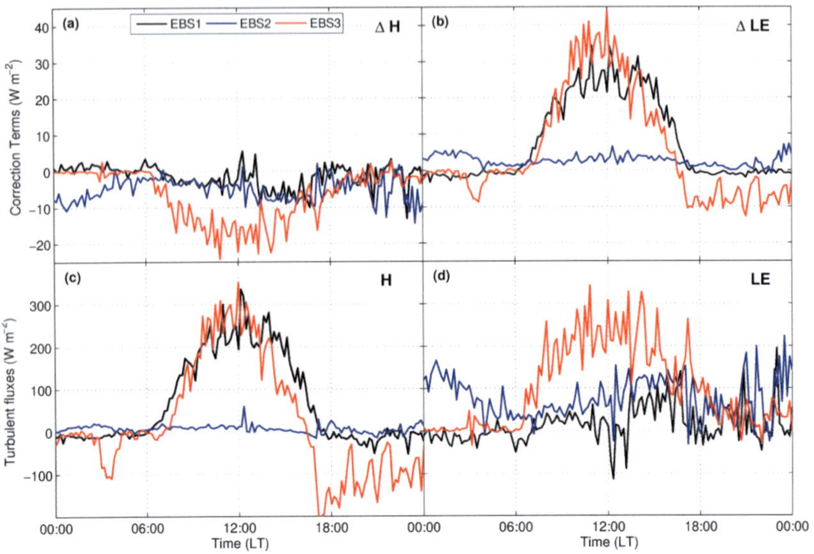

Figure 4.3: Differences between the sensible heat flux and the buoyancy flux (a), and differences between the WPL corrected and the uncorrected latent heat fluxes (b). The corresponding values for the sensible (c) and latent heat flux (d) are shown for all three measurement sites for 15 August 2014.

Webb et al. (1980) suggested the calculation of the water vapour flux (F_v) after

$$F_v = (1+\mu\sigma)\cdot\left(\overline{w'\rho_v'}+\overline{\rho_v}\cdot\frac{\overline{w'T_a'}}{\overline{T_a'}}\right), \tag{4.3}$$

with the ratio of molar masses of dry air and water vapour, $\mu = m_d/m_v \approx 1.6$, and the ratio of the water vapour to dry air density $\sigma = \overline{\rho_v}/\overline{\rho_d}$.

Fig. 4.3 (b) shows the difference between the corrected and the originally measured latent heat flux. Similar to the SND-correction, the importance of the WPL-correction varies from site to site. At EBS1 and EBS3 the correction term reaches values of up to 45 W m^{-2}, which is especially substantial at EBS1. Over water the necessary correction is relatively small with values between 0 and 10 W m^{-2}.

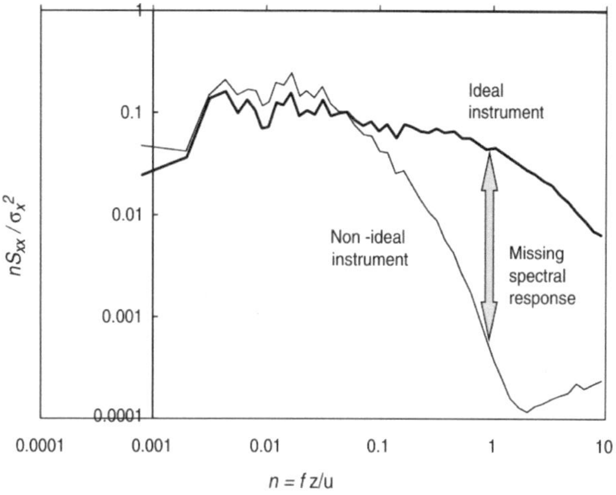

Figure 4.4: Illustration of the loss of energy for high frequencies (Foken et al., 2012b). Shown are the normalised spectra for an ideal and a non-ideal instrument. The difference between the spectra represents the missing energy which must be corrected (n normalised frequency, S_{xx} energy density of the parameter x, σ_x^2 variance of the parameter x, f frequency, z height, u wind velocity).

Spectral Corrections

The limited sensor frequency response, path-length averaging and signal processing result in a loss of energy for high frequencies. Figure 4.4 illustrates this problem. For low frequencies all energy is measured, but towards larger frequencies (small eddies) the measured spectrum varies considerably from the ideal spectrum. For open path instruments, such as the IRGASON, the frequency loss is normally small with only 5 to 10 % (Burba, 2013) but it should still be accounted for in the post-processing. The relative error on the turbulent fluxes due to spectral loss can be described by

$$\frac{\Delta F}{F} = 1 - \frac{\int_0^\infty T_{x(y)}(f) \cdot S_{x(y)}(f) df}{\int_0^\infty \cdot S_{x(y)}(f) df}, \tag{4.4}$$

with the specific transfer function ($T_{x(y)}$) and the theoretical (ideal) form of the spectrum ($S_{x(y)}$). The transfer functions used in this thesis are from Moore (1986), Kaimal et al. (1972) and Højstrup (1981) (see Mauder and Foken (2011)).

4.2.3 Quality Control

Steady State Test

The steady state test checks on non-steady state conditions within the averaging interval (Foken and Wichura, 1996). This is necessary, as the theoretical concept of the eddy covariance uses the Reynolds postulates, which require statistically stationary fluctuations within the averaging interval to be valid. The method compares the mean covariance of short time intervals of 5 min, $\overline{(x'w')_{SI}}$, with the covariance of the whole averaging interval of 30 min, $\overline{(x'w')_{WI}}$. Foken and Wichura (1996) suggested that a time interval is in a steady state if the covariance of the short interval varies less than 30 % from the covariance of the long interval. This can be expressed mathematically with

$$RN_{cov} = \left| \frac{\overline{(x'w')_{SI}} - \overline{(x'w')_{WI}}}{\overline{(x'w')_{WI}}} \right| . \quad (4.5)$$

Integral Turbulence Characteristics Test

The eddy covariance method describes the heat and mass transfer between the Earth's surface and the atmosphere as the covariance of the turbulent fluctuations of the vertical wind and a specific quantity, e.g. water vapour or temperature, and assumes fully developed turbulent conditions (Foken et al., 2012b). Therefore, the data are tested for turbulent conditions using the flux-variance similarity (Foken et al., 2012b). The flux-variance similarity states that the normalised standard deviations of turbulent parameters are constant or depend only on stability (Foken, 2006). This normalised standard deviations of turbulent parameters are also called integral turbulence characteristics (ITC). A detailed description of those ITC can be found in boundary layer or micrometeorology textbooks (Stull, 1988; Arya, 2001; Foken, 2006). With the assumption that the ITC are constant or depend only on stability, the measured integral turbulence characteristics can be compared to modelled characteristics which are taken from Foken et al. (1991) and Thomas and Foken (2002). The coefficients of these modelled characteristics can be found in Foken et al. (2012b). Again, good data quality is assumed when the deviation of the measured characteristics from the modelled characteristics is less than 30 %. The mathematical formula is

$$ITC_\sigma = \left| \frac{\left(\frac{\sigma_\chi}{\chi_*}\right)_{model} - \left(\frac{\sigma_\chi}{\chi_*}\right)_{measured}}{\left(\frac{\sigma_\chi}{\chi_*}\right)_{model}} \right|, \tag{4.6}$$

where σ_χ stands for the standard deviation of one of the wind components (u, v, w) or temperature, and χ_* is the norming parameter which is the friction velocity, u_*, for the wind components and T_* for temperature.

Flag System

Using Eqs. 4.5 and 4.6 each data point can be flagged by its quality. A quality flag between 1 and 9 refers to a certain range of ITC_σ and RN_{cov} values (see Tab. 4.2), and flag 9 is also used when data are not available for calculations. Foken (1999) suggested an overall quality flag for each data point by combining the two single data flags. The scheme he proposed is shown in Tab. 4.3, together with the percentage of sensible and latent heat flux data flagged respectively at each of the EBSs. Data from classes 1 to 3 are generally of very high quality and after Foken (1999) can be used for fundamental research. Data of the classes 4 to 6 have still a good quality and can be used for long-term measurements without limitations, whereas class 7 and 8 are only for orientation and should, together with class 9, not be used. At EBS1 the amount of data flagged as unusable (classes 7-9) was especially high. 22 % of the sensible and 37 % of the latent heat flux data were rejected (Tab. 4.3). A large amount of the rejected data could be explained by a malfunction of the IRGASON, which began on

Table 4.2: Classification of the data quality by the steady state test and the integral turbulence characteristics test. Given are the deviations of the mean covariance of a short time interval from the mean covariance of a longer averaging interval (RN_{cov}) and the deviations of the measured ITC from the modelled ITC (ITC_{σ}) after Foken (1999).

Class	Difference Range (%)
1	0 - 15
2	16 - 30
3	31 - 50
4	51 - 75
5	76 - 100
6	101 - 250
7	251 - 500
8	501 - 1000
9	> 1000 / not available

11 October 2014. At EBS2 and EBS3 data availability was better and the first group (class 1-3) accounted for over 60 % of the flux data and data amounts rejected were 16 % or less (Tab. 4.3).

4.3 Assessment of Missing Parameters

4.3.1 Latent Heat of Vaporisation Determined for Dead Sea Water

The latent heat of vaporisation is needed to calculate the latent heat flux of the Dead Sea water surface from the IRGASON measurements (see Sec. 2.3 and 3.3). It depends on salinity and temperature. Many studies were carried out to determine the properties of sea salt solutions and their dependency on temperature and salinity. As the studies are based on experimental data, the found correlations are only valid for a given temperature and salinity range. Many studies investigated the typical oceanographic range with maximum salinities

Table 4.3: Combination of the single quality flags into an overall quality flag after Foken (1999), and the percentage of the sensible and latent heat flux data from the three EBSs flagged respectively.

Overall flag	Steady State Test	Integral Turbulence Characteristics Test	EBS1		EBS2		EBS3	
			H (%)	LE (%)	H (%)	LE (%)	H (%)	LE (%)
1	1	1 - 2	41.9	14.5	41.7	45.4	47.0	42.0
2	2	1 - 2	8.4	9.4	12.6	12.2	6.3	7.8
3	1 - 2	3 - 4	5.2	3.0	8.3	10.1	16.0	11.2
4	3 - 4	1 - 2	9.0	17.0	10.7	7.9	4.8	5.2
5	1 - 4	3 - 5	4.3	4.0	7.6	5.9	5.8	7.5
6	5	≤ 5	2.4	7.7	5.4	3.6	3.0	3.1
7	≤ 6	≤ 6	4.3	12.6	7.5	5.2	2.8	4.8
8	≤ 8	≤ 8	2.2	5.2	3.2	2.5	2.0	2.2
9	one of the flags equal to 9		15.1*	19.4+	2.9	7.2	5.3	9.3x
Instrument dismounted in February			7.1	7.1	0.0	0.0	6.8	6.8

*12.1 % due to instrument failure beginning in October 2014
+16.5 % due to instrument failure beginning in October 2014
x4.3 % due to instrument failure beginning in January 2015

of about 40 g kg^{-1}. Others considered higher salinities up to 120 g kg^{-1} and temperatures from 0 to 200 °C, as this is the relevant range for desalination applications, but no studies were carried out for salinities as high as they are found for the Dead Sea ($S \approx 277$ g kg^{-1}; Lensky et al. (2005)). The only available data for Dead Sea water were constant values of the latent heat of vaporisation at 22 or 25 °C. The surface water temperature of the Dead Sea varies considerably over the course of a day and also over the course of the year. Hence, these constant values of the latent heat of vaporisation are not applicable for the calculation of the latent heat flux with a temporal resolution of 30 min. Therefore, laboratory measurements to determine the temperature dependency of the latent heat of vaporisation were performed (experimental setup see Sec. 3.3).

First, the experiment was conducted with pure water and the saturation vapour pressure of pure water, E_w, was measured as a function of water temperature.

The results were compared to literature values and a linear regression to literature values from the Kilolabor ETH Zurich[1] was calculated. Afterwards, the saturation vapour pressure of the saline water, E_{sw}, was measured as a function of water temperature. The linear regression formula of E_w, obtained using the literature values, was used to correct the E_{sw} values. Through this approach possible measurement uncertainties of the manometer can be minimised. From these two experiments the activity of water can then be calculated as:

$$a_{sw} = \frac{E_{sw}}{E_w}. \tag{4.7}$$

The averaged activity for the Dead Sea water is $a_{sw} = 0.65$.
The molar latent heat of vaporisation, ΔH_v in J mol^{-1}, is then derived by using the general form of the Clausius-Clapeyron equation, assuming that the molar volume of the liquid can be neglected against the molar volume of the gas, and by using the ideal gas law:

$$\Delta H_v = -R\,\frac{\mathrm{d}(\ln E_{sw})}{\mathrm{d}(\frac{1}{T_w})}. \tag{4.8}$$

R=8.314 J mol^{-1} K^{-1} is the universal gas constant. With the molar mass of water, $m_{\mathrm{H_2O}}$=0.018 kg mol^{-1}, the specific latent heat of vaporisation, L_v, can be calculated as follows:

$$L_v = \frac{\Delta H_v}{m_{H_2O} \cdot 1000}, \tag{4.9}$$

in kJ kg^{-1}, and can then be fitted to the water temperature (Fig. 4.5). The regression formula is:

$$L_v = 5150.6561 - 13.9530 \cdot T + 0.0162 \cdot T^2, \tag{4.10}$$

with L_v in kJ kg^{-1} and T in K. Eq. 4.10 will be used in this thesis to calculate the latent heat flux of the Dead Sea water surface from the flux measurements.

[1] https://cdm.unfccc.int/filestorage/U/4/B/U4BKYDK7NTLWWFQ1OTUFUCKJMTEE3Y/U4BKYDK7.pdf?t=Vm98bzQ0aGx1fDC3cDweIA5PuHui7yRAOy3k

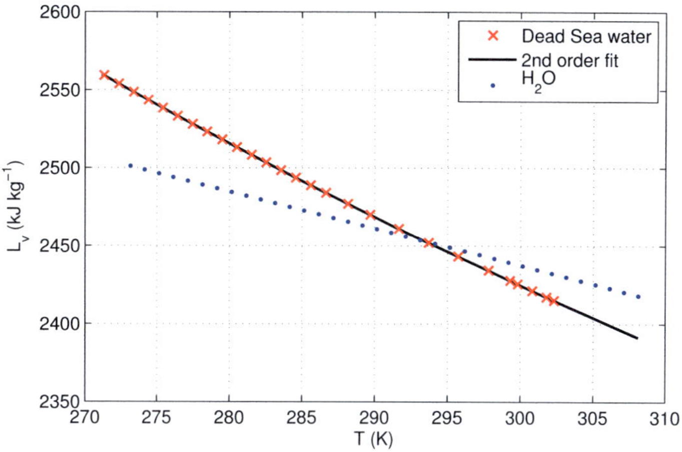

Figure 4.5: Dependency of the specific latent heat of vaporisation (L_v) on temperature. Laboratory measurements of L_v of the saline water of the Dead Sea, a second order polynomial fit, and literature values for pure water (H_2O) are shown.

4.3.2 Calculation of Vapour Pressure Deficit

The vapour pressure deficit (Δe) is one of the factors controlling the intensity of the latent heat flux from a water surface. In Sec. 6.2 the dependency of the latent heat flux of the Dead Sea water surface on the vapour pressure deficit will be determined. Additionally, the measurements will be compared with the results of used evaporation formulas which also use the vapour pressure deficit as an input variable (Sec. 2.3.2). Therefore, a short explanation of the calculation of the vapour pressure deficit will be given here. The vapour pressure deficit over a water surface is defined as the difference between the saturation vapour pressure above the water (E_w) and the atmospheric vapour pressure (e_a)

$$\Delta e = E_w - e_a. \tag{4.11}$$

For the Dead Sea the salinity of the lake has to be considered when calculating the saturation vapour pressure of the saline water (E_{sw}), as it is lower for saline water than it is for freshwater by a factor a_{sw}, caused by the vapour pressure depression resulting from the dissolved salts (Raoult's law) (Atkins, 2014),

$$E_{sw} = a_{sw} \cdot E_w. \tag{4.12}$$

The activity of the saline water, a_{sw}, depends on the composition of the dissolved salts, and its value is 0.65 for the Dead Sea water (see Sec. 4.3.1). The saturation vapour pressure over freshwater depends only on the surface water temperature (T_s) and can be calculated using the Magnus equation after Bolton (1980):

$$E_w(T_s) = 6.112 \cdot \exp\left(\frac{17.67 \cdot (T_s - 273.15)}{T_s - 29.65}\right), \tag{4.13}$$

with E_w in hPa. As surface water temperature is not directly measured at EBS2, two approaches are applied:

1. Vapour pressure deficit is calculated using water temperature at 1 m depth, $T_{w,1\mathrm{m}}$ in K, instead of surface water temperature:

$$\Delta e_{T_w} = e_a - a_{sw} \cdot E_w(T_{w,1\mathrm{m}}). \tag{4.14}$$

2. Vapour pressure deficit is calculated using surface water temperature obtained by a special application of the Monin-Obukhov theory, T_{MO} in K:

$$\Delta e_{MO} = e_a - a_{sw} \cdot E_w(T_{MO}), \tag{4.15}$$

   where T_{MO} is calculated with the following profile equation:

$$T_{MO} = T(z_0) = T(z_m) - \frac{\Theta_*}{\kappa} \cdot \left(\ln\frac{z_m}{z_0} - \Psi_H(\zeta_m, \zeta_0)\right). \tag{4.16}$$

$\zeta_m = z_m L_*^{-1}$ and $\zeta_0 = z_0 L_*^{-1}$ are independent dimensionless parameters using the Monin-Obukhov Length, L_*, and $\frac{\Theta^*}{\kappa}$ is a scaling parameter defined as:

$$\frac{\Theta_*}{\kappa} = -\frac{1}{\kappa u_*} \frac{H}{\rho_a c_p}. \tag{4.17}$$

Ψ_H is the integral over the empirically determined functions φ_H:

$$\Psi_H(\zeta_m, \zeta_0) = \int_{\zeta_0}^{\zeta_m} = \frac{1 - \varphi_H}{\zeta} d\zeta. \tag{4.18}$$

In this work the φ functions from Dyer (1974) are used:

$$\varphi_H = 1 + 5\zeta \qquad \zeta > 0 \tag{4.19}$$

$$\varphi_H = (1 - 16\zeta)^{-1/2} \qquad -1 < \zeta < 0. \tag{4.20}$$

The differences between the two approaches to calculate the vapour pressure deficit can best be seen in winter. Exemplary diurnal cycles of the vapour pressure deficits calculated using these two approaches, the respective surface water temperature, water temperature in 1 m depth, and the air temperature are shown in Fig. 4.6. In winter, the water temperature at 1 m depth (blue line) is nearly constant throughout the day and amounts to 25 °C. On the other hand, the air temperature (black line) has a clear diurnal cycle but is lower than the water temperature. Especially during the night the difference is about 5 °C. The surface water temperature calculated using approach II (red line) does also have a diurnal cycle, even if it is small. During the night it is about 21 °C, and during the day it reaches 24 °C. This leads to a considerably smaller vapour pressure deficit calculated using approach II (red line) compared to the one using approach I. Additionally, the diurnal cycles are reversed. Approach I has higher values during the night and lower values during the day and for approach II it is vice versa. This leads to a difference between the two calculated vapour pressure deficits of about 5 hPa during the night and 1 hPa during the day.

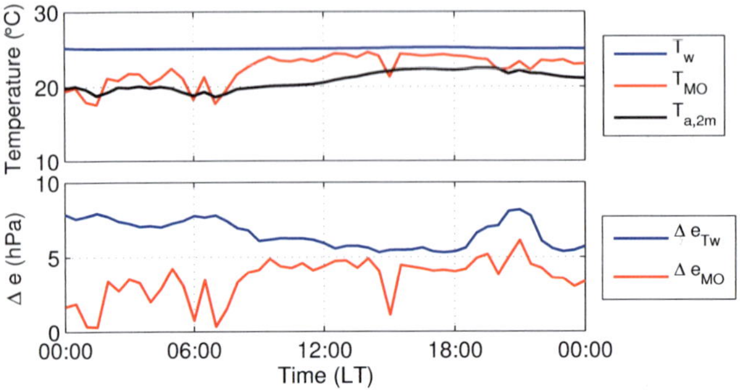

Figure 4.6: Diurnal cycles of water temperature in 1 m depth (T_w), surface water temperature (T_{MO}), 2 m air temperature (T_a), vapour pressure deficit calculated with T_w (Δe_{T_w}) and vapour pressure deficit calculated with T_{MO} (Δe_{MO}), on 29 November 2014.

4.3.3 Calculation of the Net Radiation of the Water Surface

EBS2 is located at the shoreline in order to measure the conditions of the water surface. This works well for measurements of temperature, humidity and the turbulent fluxes, but not for the calculation of the net radiation. The measurements of the radiation components of the lower half space are not conducted directly over the water surface, but over the land surface. Therefore, these two components have to be calculated. The longwave upward radiation is proportional to the fourth power of the surface water temperature and can be calculated using the Stefan-Boltzmann equation (Eq. 2.15) and the surface water temperature derived in Sec. 4.3.2, T_{MO}. For the emissivity of a water surface $\varepsilon = 0.98$, a common literature value (e.g. Konda et al. (1994)) is used. For the calculation of the shortwave upward radiation, the albedo of the water surface is needed. Therefore, a short-term experiment close to EBS2 was conducted to derive the albedo from the water surface near the station. The experimental setup is explained in Sec. 3.2.1. The albedo was only calculated between 10 and 14 LT as the albedo depends on the solar elevation angle and is constant only for high elevation angles during midday. Calculations were performed for the

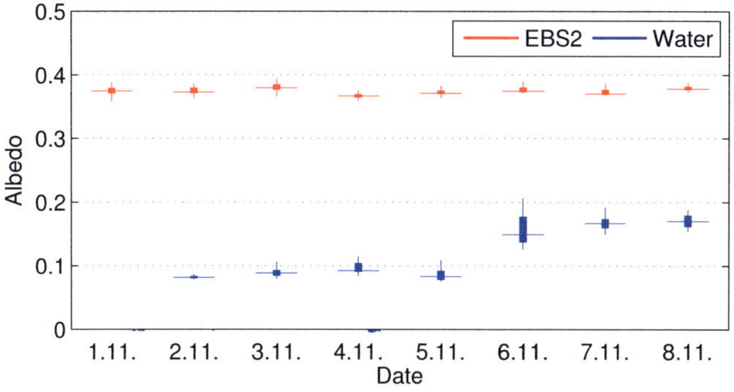

Figure 4.7: Boxplot for albedo values, derived between 10 and 14 LT for the land surface at EBS2 and for the water surface close to EBS2 from the 1 until 8 November 2014. The horizontal lines mark medians, the edges of the boxes are the 25th and 75th percentiles and whiskers extend to the most extreme data points.

measurements above the water surface and for the measurements undertaken at EBS2. As the albedo is related to the surface it should be constant as long as the underlying surface does not change. This was the case for the measurements at EBS2. The median of the albedo measurements varied between 0.37 and 0.38 (Fig. 4.7). On the other hand, for the measurements over water, the albedo was only constant at the beginning of the measurements with values between 0.08 and 0.09, and then suddenly jumped to 0.14 and 0.16 (Fig. 4.7). This is most likely due to salt accumulation at the instrument caused by the sea spray, which can also be seen in Fig. 3.5 c. The last three days should therefore not be considered, resulting in an average albedo of 0.08. Values found in the literature support this measurements. Stanhill (1987) calculated the albedo of the Dead Sea surface from ship measurements and reports values of 0.06 in the summer months and 0.09 in the winter months and an annual average of 0.07. He also reported albedo values from Kondrat'Ev (1969) for the latitude of the Dead Sea and the cloud cover observed in the northern part of the Dead Sea, which was also 0.08 for November and 0.07 as an average annual albedo value. Thus, for the calculation of the net radiation in this work an albedo of 0.07 is used to account for lower albedo values in summer compared to those measured in November. The net radiation at the water surface is so calculated using

$$Rn = (1 - 0.07) \cdot K\downarrow + L\downarrow - 0.98 \cdot \sigma \cdot T_{MO}^4. \tag{4.21}$$

4.4 Stepwise Linear Regression Model

In Sec. 6.2, a regression model will be applied to find a relationship between the turbulent fluxes from the Dead Sea water and several of the meteorological parameters, such as vapour pressure deficit, water temperature, surface water temperature, and wind speed. The general work flow applied to determine this relationship will be explained here. The data used are measured at EBS2. The surrounding area of EBS2 has two main land cover types: water and bare soil (Fig. 3.4 b). To find a relationship for the turbulent fluxes from the water surface, only flux data, where the fetch is over water, can be used for the model. Therefore, all flux data for wind directions between 230° and 330° were removed from the data set. To avoid excessive influence of outliers on the regression calculations, all data exceeding the 99th percentile or falling below the 1st percentile were excluded from the data set additionally. The remaining data were then used with a stepwise linear regression model. The stepwise linear regression model uses a systematic approach to find independent variables (e.g. vapour pressure deficit, wind speed, temperature) explaining the desired dependent variable (sensible or latent heat flux). The model uses a forward and backward algorithm which checks at each step if a term can be added or removed from the model (Draper and Smith, 1998). The algorithm starts with a constant model and gradually adds a term if it results in the increase of the coefficient of determination (R^2) by at least 0.05. In the same way a term is removed from the model if it results in a decrease of R^2 by less than 0.04. To test the model on its robustness, a cross-validation can be performed (Stone, 1974; Xu et al., 2004). The general goal of such a cross-validation is the definition of a data set to build the model and a validation data set with which the model is tested.

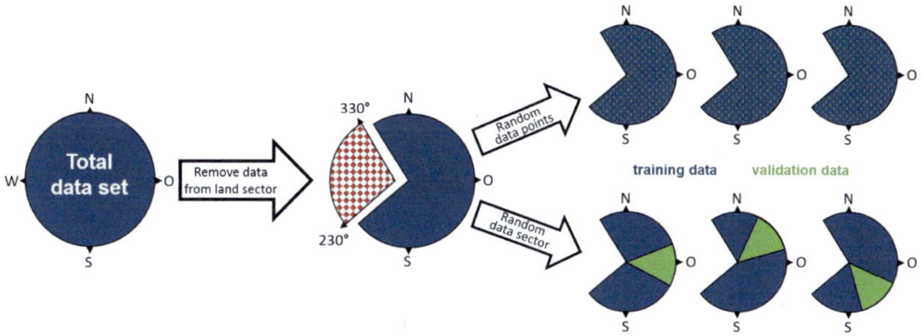

Figure 4.8: Schematic of the work flow of the Monte-Carlo cross validation.

The cross-validation method used in this work is the so called Monte-Carlo cross validation (MCCV), first introduced by Picard and Cook (1984), which randomly divides the data in a training and validation data set. A schematic of the work flow is shown in Fig. 4.8. First, data with a wind direction between 230° and 330° are removed from the data set, then two approaches to split the remaining data are used:

1. The validation data set consists of randomly chosen data points of approximately 15% of the whole data set.
2. The validation data set consists of the data of a randomly chosen wind direction sector of 45°.

After each split a regression model for the turbulent flux is build from the training data, and the values for the validation data set are calculated with the model. The deviation of the estimated values from the measured values of the validation data set yield the model error of one realisation. After multiple rounds of splitting the data set the results can be averaged. The prediction error (*er*) of the regression model is then the mean of the model errors from all realisations. The prediction error shows if the model is independent of the data used to build the model and if the results are robust. A large prediction error indicates that the model depends on the choice of the training data set and is therefore not applicable for new data sets with which the model is not trained on. In case of the flux data from EBS2, a large model error would indicate that the found model is not able to correctly estimate the fluxes from the land sector (230 - 330°).

5 Wind Systems in the Valley

To characterise the typical wind systems in the Dead Sea valley and to investigate their frequency of occurrence, long-term wind measurements of the energy balance stations (EBSs) from March 2014 until March 2015 were used. Additionally, case study data from the SOPs were used to investigate the details of the wind systems and their respective trigger mechanisms and to analyse their impact on the atmospheric conditions in the valley in more detail. The intra-annual variation and frequency of the wind systems and their general characteristics are described in Sec. 5.1, followed by a case study investigating the typical wind systems and trigger mechanisms in summer (Sec. 5.2). In Sec. 5.3 the interaction of a strong large-scale flow with the complex orography of the Dead Sea region are presented using a case study of the winter SOP. The results are discussed in Sec. 5.4.

5.1 Diurnal Variation of Wind Systems

For the analysis of the diurnal wind variations also the intra-annual variability has to be considered. Therefore, the analysis of the wind systems occurring in the Dead Sea valley was performed for each meteorological season separately using the diurnal cycle of the relative frequency of the wind directions at EBS1 (Fig. 5.1). To account for the high variability of the wind field, a threshold of the relative frequency of 0.05 was used to identify the main patterns (black contours in Fig. 5.1). Several distinct patterns can be identified. During daytime a north-easterly flow or a south-easterly flow occurred, which are afterwards referred to as P1 and P2. In the evening a westerly flow was observed (P3a) and another westerly maximum was found during the night (P3b). Finally, a northerly flow, which always occurred after 20 LT, was identified as P4. For each pattern all relative frequency values above the threshold were accumulated at each time step, resulting in time series of the total frequency of occurrence (tfo) of each pattern (Fig. 5.1, lower panels).

5.1.1 Lake Breeze

The first pattern (P1), representing a north-easterly wind, occurs in all seasons during daytime. In summer it is observed on over 70 % of the days (Fig. 5.1 b), and in spring and autumn it accounts for at least 60 % and 50 % of the daytime conditions, respectively (Fig. 5.1 a and c). In winter this pattern only occurs on 25 % of the days. Besides P1 there is a second pattern (P2) with a south-easterly flow during daytime, which can only be observed in winter. P2 accounts for another 20 % of daytime wind conditions in winter (Fig. 5.1 d).
P1 develops shortly after sunrise, between 06:00 and 08:00 LT depending on the season. It develops when the difference between the sensible heat flux of the land surface and the water surface becomes positive caused by a stronger heating of the near-surface air mass above the land than over the water (Fig. 5.2). This horizontal temperature gradient, results in a horizontal pressure gradient and leads to the formation of a lake breeze. The diurnal cycles of the wind direction show a shift of the median wind direction from west to east, and median wind speed increases to about 3 $m\,s^{-1}$ (Fig. 5.2). In summer the variation in wind speed during daytime is very small, as can be seen from the 25th and 75th percentile of the diurnal cycle of the wind speed (Fig. 5.2 b, shaded area), and can be attributed to the very weak synoptic forcing throughout the summer season. The weak synoptic forcing results in very persistent and similar atmospheric conditions in the valley on each day in summer causing a similar temperature gradient between the water and land surface and therefore resulting in very consistent lake breeze conditions. The same applies to spring and autumn; only in winter a larger variation in wind speed can be seen, which can be explained by the stronger influence of the synoptic conditions and the

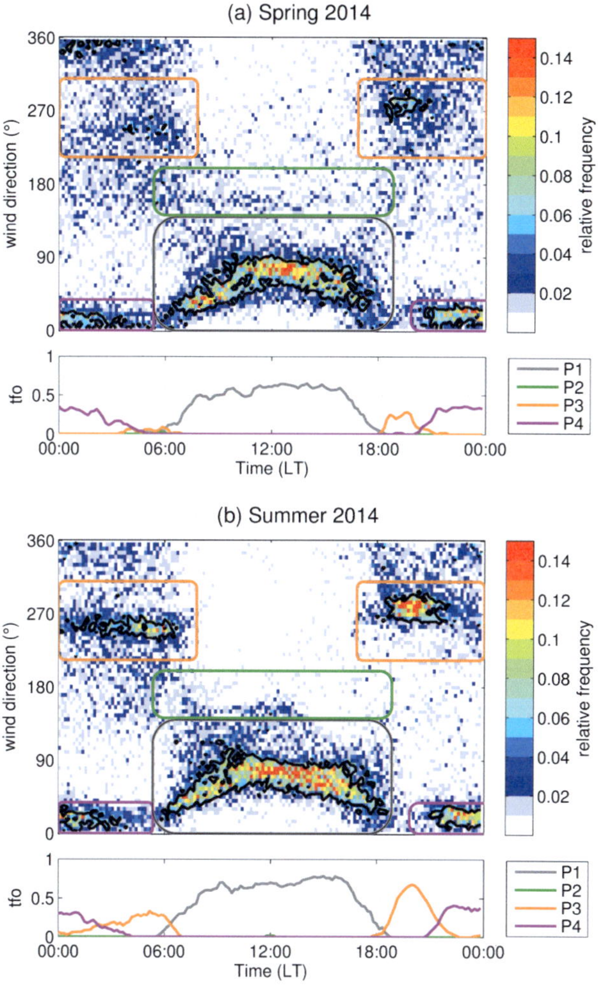

Figure 5.1: Diurnal cycle of the relative frequency of wind direction (top panels) for spring (a) and summer (b). Marked with the black contour is the 0.05 threshold. The diurnal cycle of the total frequency of occurrence (tfo) above the 0.05 threshold for the different patterns is shown in the bottom panels for each season. P1 is the north-easterly lake breeze, P2 is caused by a superposition of the lake breeze and the large-scale flow, P3 are the westerly DWFs, and P4 are nocturnal northerly along-valley flows.

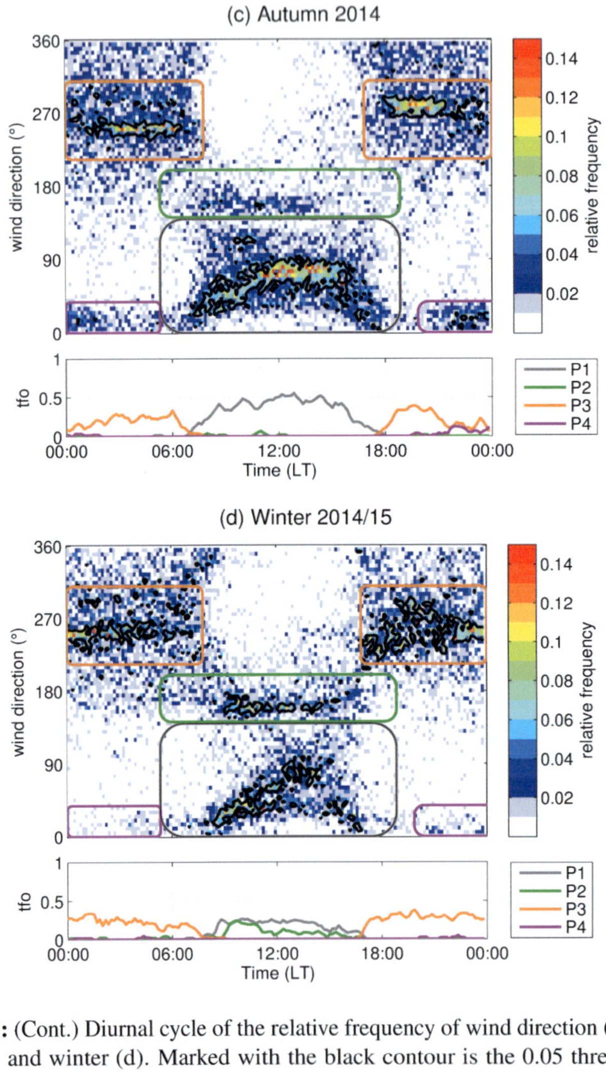

Figure 5.1: (Cont.) Diurnal cycle of the relative frequency of wind direction (top panels) for fall (c) and winter (d). Marked with the black contour is the 0.05 threshold. The diurnal cycle of the total frequency of occurrence (tfo) above the 0.05 threshold for the different patterns is shown in the bottom panels for each season. P1 is the north-easterly lake breeze, P2 is caused by a superposition of the lake breeze and the large-scale flow, P3 are the westerly DWFs, and P4 are nocturnal northerly along-valley flows.

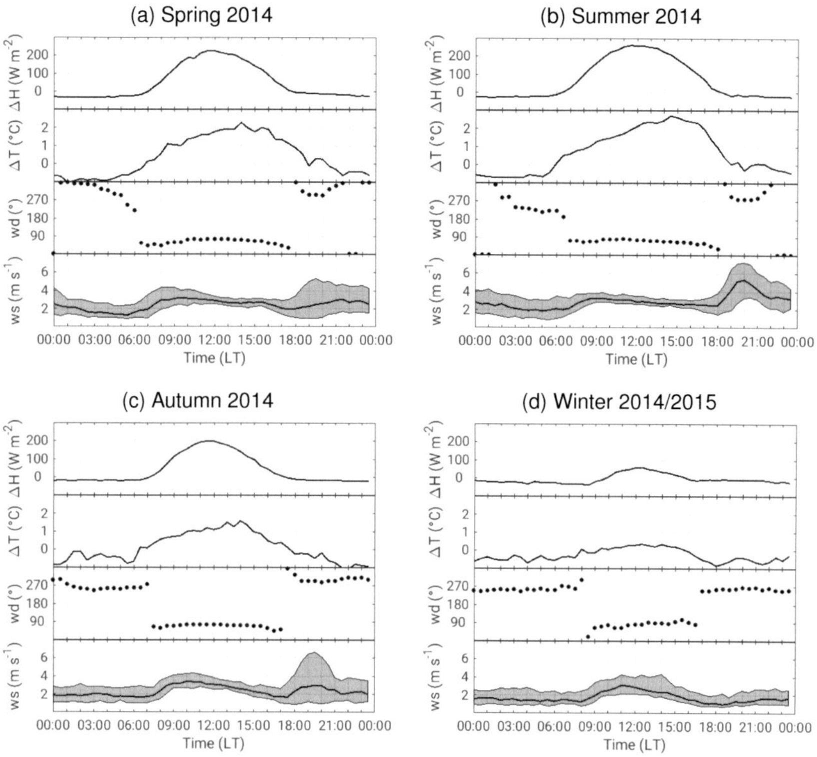

Figure 5.2: Diurnal cycle of the median sensible heat flux difference (Δ H) and temperature difference (Δ T) between land (EBS1) and water surface (EBS2), and diurnal cycle of median wind direction and wind speed over land (EBS1). For the wind speed the 25th and 75th percentiles are shown as shaded layer. Data are shown for spring 2014 (a), summer 2014 (b), autumn 2014 (c), and winter 2014/2015 (d).

atmospheric conditions in the valley in winter (Fig. 5.2 d). The lake breeze persists during daytime until the sensible heat flux difference between land and water becomes negative again and other wind patterns prevail. P2 represents a south-easterly flow and only appears in winter, which indicates the influence of the large-scale pressure gradient on the atmospheric conditions in the valley in this season. A superposition of the north-easterly lake breeze with large-scale forcing, e.g. a Red Sea Trough with an western axis, causes a shift to south-easterly wind directions.

The lake breeze has certainly a large influence on the atmospheric conditions in the valley as it is the predominant feature in the wind field during daytime in all seasons.

5.1.2 Westerly Flow

P3 represents westerly wind conditions and can be subdivided into: P3a, a pronounced maximum of westerly wind in the evening in spring, summer, and autumn, and P3b, a secondary longer lasting westerly flow during the rest of the night, which is observed in summer, autumn and winter (Fig. 5.1).

P3a develops in the evening soon after the lake breeze ceases. The mean wind direction changes suddenly to west at around 17:30 to 18:00 LT for three to four hours (Fig. 5.1). This change can be observed in spring, summer and autumn, but in spring and autumn it appears less frequently. The occurrence is about 65 % in summer, but only 28 % in spring and 35 % in autumn (Fig. 5.1). For the other days with no westerly flow, especially in winter, wind direction varies from south-westerly to northerly flow with no dominant pattern (Fig. 5.1). Also wind speed varies strongly during this time interval in the evening. Median wind speed is about $5\,\mathrm{m\,s^{-1}}$ in summer, and the 25th and 75th percentiles are already $2\,\mathrm{m\,s^{-1}}$ below and above the median (Fig. 5.2 b). In spring and autumn, median wind speed is only $3\,\mathrm{m\,s^{-1}}$, and the 25th and 75th percentile are 1 and $7\,\mathrm{m\,s^{-1}}$, respectively. For a precise analysis of the wind velocities related to the west wind events, wind roses were calculated for the time between 17:30 and 21:30 LT for spring, summer, and autumn (Fig. 5.3). The analysis shows that the westerly flow causes high wind speeds with values ranging from 5 up to $12\,\mathrm{m\,s^{-1}}$. The wind velocities for the other occurring wind directions were considerably lower, mainly below $3\,\mathrm{m\,s^{-1}}$. As the station is located at the western shore of the Dead Sea not more than 600 m away from the mountains, these Westerlies can be attributed to DWFs from the Judean Mountains.

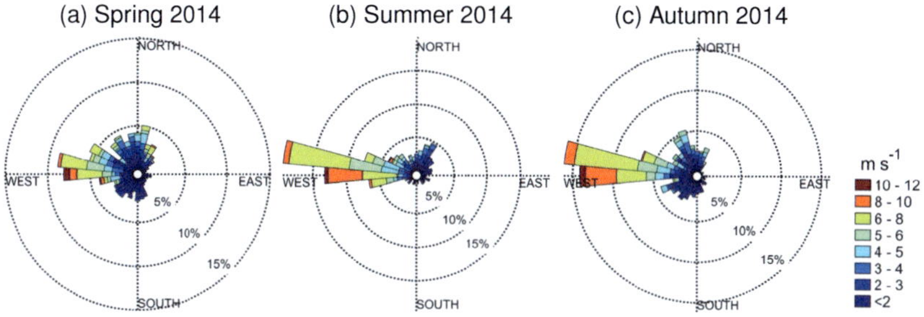

Figure 5.3: Evening wind conditions at EBS1 between 17:30 and 20:30 LT. Data are shown for spring (a), summer (b), and autumn 2014 (c).

Several of these DWFs were observed during the first SOP in August 2014. On 15 days strong westerly DWFs were observed east of Masada in the evening. For 10 of those 15 events lidar data were available. The vertical profiles of the horizontal wind calculated with the VAD algorithm after Browning and Wexler (1968) are shown in Fig. 5.4 for three of these days. It can be observed that the characteristics of the westerly DWFs are rather diverse with regard to the height of the west wind layer, the wind speed, and the duration of the event. Fig. 5.4 (a) shows a short and weak event which lasted only about 2 h from 18:00 to 20:00 LT, and maximum wind speed was around 6 m s^{-1}. The second example shows a very strong event which lasted about 4:30 h and wind speeds exceeded 10 m s^{-1}. The last example again shows a rather weak westerly flow of about 6 m s^{-1}, but it lasted over the whole night. With the vertical profiles of the horizontal wind speed gained from the VAD algorithm it is difficult to specify the characteristics of the west wind layer in more detail. Therefore, the systems were analysed using lidar RHI scans at 299°. Averaged vertical profiles of the horizontal wind component were calculated and used for the further analysis. A detailed description of the method can be found in Sec. 4.1. The duration, the mean wind speed and mean height of the west wind layer as well as the mean maximum wind speed and its height are shown in Tab. 5.1 for each of the 10 events.

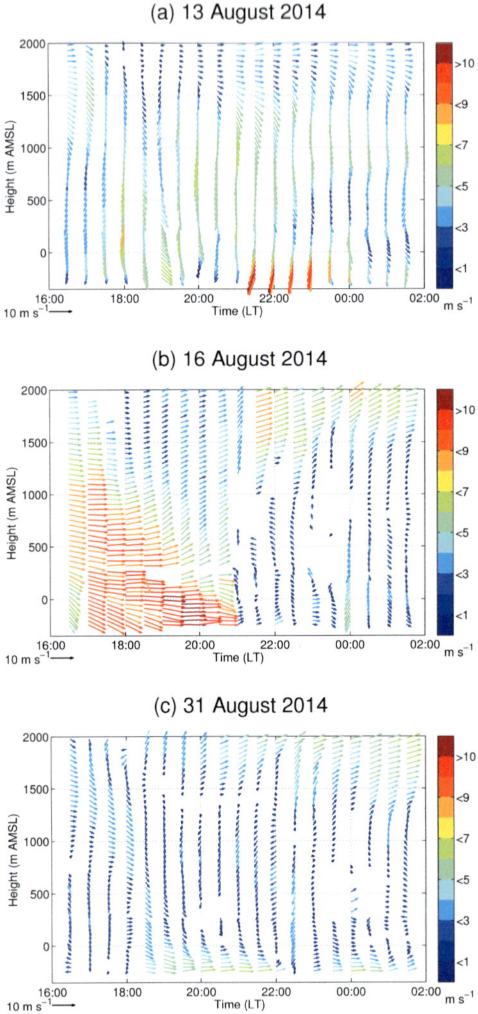

Figure 5.4: Horizontal wind vectors calculated with the VAD algorithm from lidar data at the KITcube site near Masada (Fig. 3.6). For the lowest 500 m Windcube data were used and for the layers above data from the Windtracer. Length and colours of the arrows represent wind velocity and the direction of the arrows indicate the wind direction. 10 min mean profiles are shown for a (a) short, (b) medium, and (c) long lasting DWF.

Table 5.1: Characteristics of the DWF events derived from RHI scans at 299°. Presented are mean wind speed of the layer, mean height of the DWF, mean maximum wind speed, and mean height of the wind speed maximum. The mean wind speed was first calculated for each time step and then averaged over the whole event. The height, maximum wind speed, and height of the wind speed maximum were determined for each time step and then also averaged over the whole duration of the event. The events are grouped according to their duration.

Date	mean wind speed ($m\,s^{-1}$)	mean layer height (m AGL)	mean maximum wind speed ($m\,s^{-1}$)	mean height of maximum wind speed (m AGL)	duration (hh:mm)
13.08.2014	3.4	420.P	5.1	199.0	1:51
22.08.2014	4.1	529.1	5.6	293.6	2:09
25.08.2014	6.2	501.8	8.3	312.7	2:39
14.08.2014	3.0	600.0	3.1	407.4	3:14
26.08.2014	7.2	961.1	10.8	406.8	4:13
15.08.2014	7.5	1338.5	10.1	574.8	5:12
16.08.2014	6.8	924.7	8.7	493.0	5:18
18.08.2014	5.1	834.4	7.5	450.9	5:24
28.08.2014	5.4	386.5	8.3	201.8	11:47
31.08.2014	4.0	251.0	6.1	124.0	11:41

The events can be grouped according to their duration in short (2 - 3 h), medium (4 - 5:30 h) and long (over 11 h) DWFs. Not only the duration is similar in the groups, but also several other characteristics and they will now be discussed for each group separately.

DWFs of Short Duration

Short DWF events were observed on 13, 14, 22, and 25 August 2014. All four cases showed the same behaviour before the west wind was observed in the valley. An elevated west wind maximum around crest height existed, which then penetrated into the valley over a time of about 2 h until the strong wind speeds reached the valley floor (Fig. 5.5). In the valley the west wind was observed at

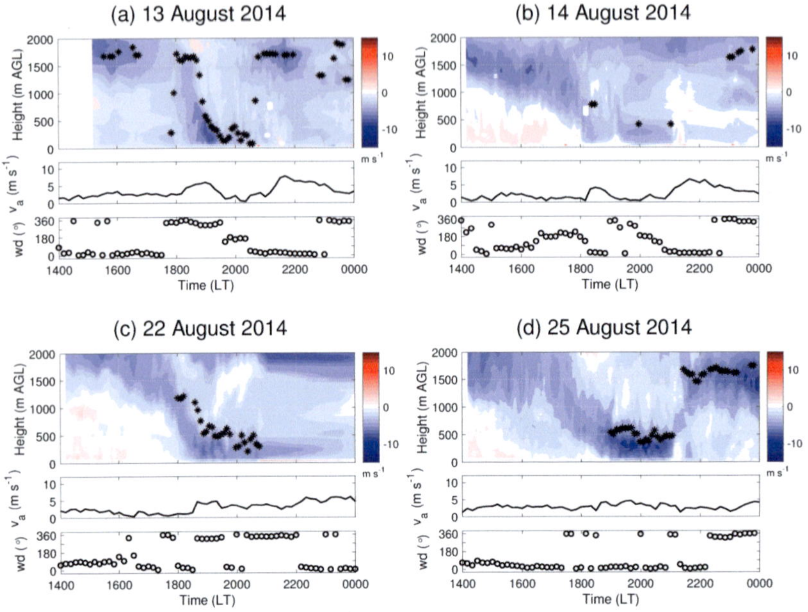

Figure 5.5: Averaged vertical profiles of the horizontal radial wind component derived from lidar RHI scans at 299°(method described in Sec. 4.1). Negative values (blue colours) indicate a wind component from north-west and positive values (red colours) indicate a wind component from south-east. Black stars indicate the detected height of the layer characterised by severe westerly flow. Below the cross sections wind direction and wind speed measurements in 40 m height measured with the small lidar (Windcube) are shown. Data are shown for four events with a duration of only 2 - 3 h.

around 18:00 to 19:00 LT. The event on 25 August 2014 was the strongest event, with a mean wind speed of 6.2 m s^{-1} (Tab. 5.1). The other three events were quite comparable in their strength, with a mean wind speed that varied between 3.4 and 4.1 m s^{-1}. The mean layer height varied between 420 and 600 m AGL, and the mean height of the wind speed maximum was between 199 and 407 m AGL. The maximum wind speed, averaged over the event duration, was obviously strongest on 25 August with 8.3 m s^{-1}, and between 3.1 and 5.6 m s^{-1} for the other three cases. To evaluate how far the DWF penetrated into the valley, Windcube

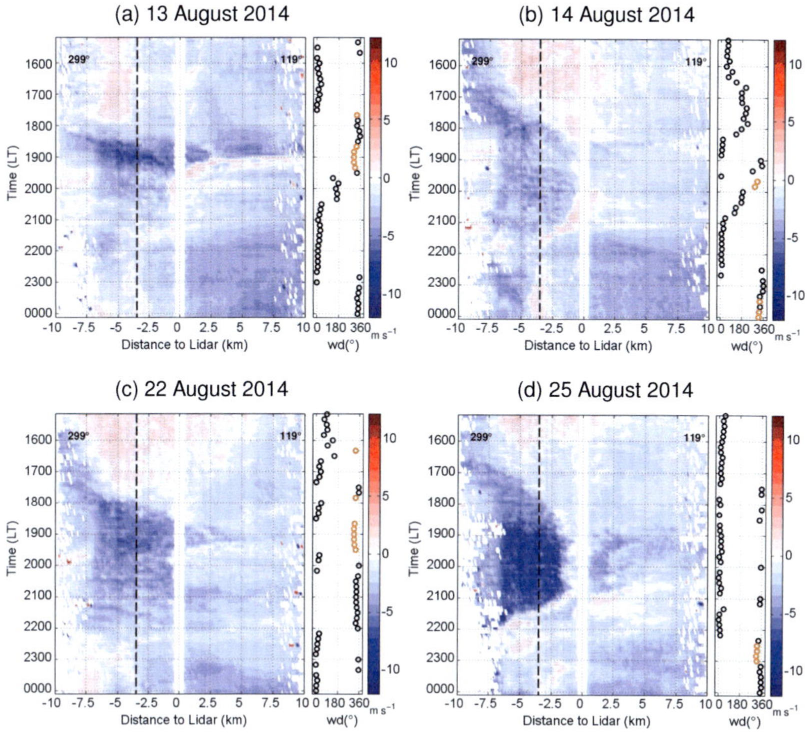

Figure 5.6: Hovmoeller diagrams of the near-surface radial wind velocity derived from lidar RHI scans at an azimuth angle of 119° and 1.6° elevation, and an azimuth angle of 299° and 7° elevation. Negative values (blue colours) indicate north-west wind and positive values (red colours) wind from south-east. The black dashed line indicates the end of the slope and the beginning of the valley floor. Additionally, wind direction measured with the small lidar system in 40 m AGL is shown. Orange points highlight wind directions between 270 and 330°, indicating that the DWF reached the measurement location. Data are shown for the four events with a duration of only 2 - 3 h.

measurements at 40 m AGL were used together with Hovmoeller diagrams. The Hovmoeller diagrams were derived from the near-surface radial velocity taken from lidar RHI scans at an azimuth angle of 119° and 1.6° elevation and at an azimuth angle of 299° and 7° elevation. The Hovmoeller diagrams (Fig. 5.6) show that between 15:00 and 18:00 LT upslope winds prevailed, which were then replaced by westerly DWFs. The west winds reached the valley floor approximately at -3.5 km north-west of the lidar but did not penetrate far into the valley (Fig. 5.6). A comparison to the measurements directly at the lidar location shows that a west wind was observed at the lidar location on 13 and 22 August for only 1 h and even less than one hour on 14 August (orange points Fig. 5.6). The DWF reached approximately 6 km into the valley, but a proper penetration distance could not be identified due to the weak wind velocities. On 25 August the west wind did not reach the lidar location at all, even though it was the strongest event. It stagnated at approximately 500 m away from the lidar location. On all four days between 21:00 and 22:00 LT the DWF was no longer observable in the valley because an along-valley flow began to dominate (Fig. 5.4 (a) and 5.7). This can also be seen in the 40 m wind direction and wind speed measurements. Especially on 13 and 14 August at 21:00 LT a strong northerly flow with wind speeds up to 7 m s^{-1} started and dominated the wind field for several hours (Fig. 5.5). On the other two days the along-valley flow was not as strong but could still be observed at the station.

DWFs of Medium Duration

This group of events had a general duration of about 4 to 5:30 h and was observed on 15, 16, 18, and 26 August. These events were observed earlier than the short events, at around 16:00 LT (Fig. 5.8). Similarly to the short events an elevated west wind maximum around crest height penetrated into the valley, but affected a deeper layer. The west wind layer height for those events was around 1200 m AGL at the beginning of the event and decreased towards the end of the event to only 500 m AGL or less on three of the four days (Fig. 5.8). Only on 15 August the west wind layer was quite thick over the whole period with 900 to 1700 m. The mean layer heights were therefore quite high with 834 up to 1338 m AGL (Tab. 5.1). The DWF layer was not only thicker but was also stronger compared to the short events. The mean wind speed was

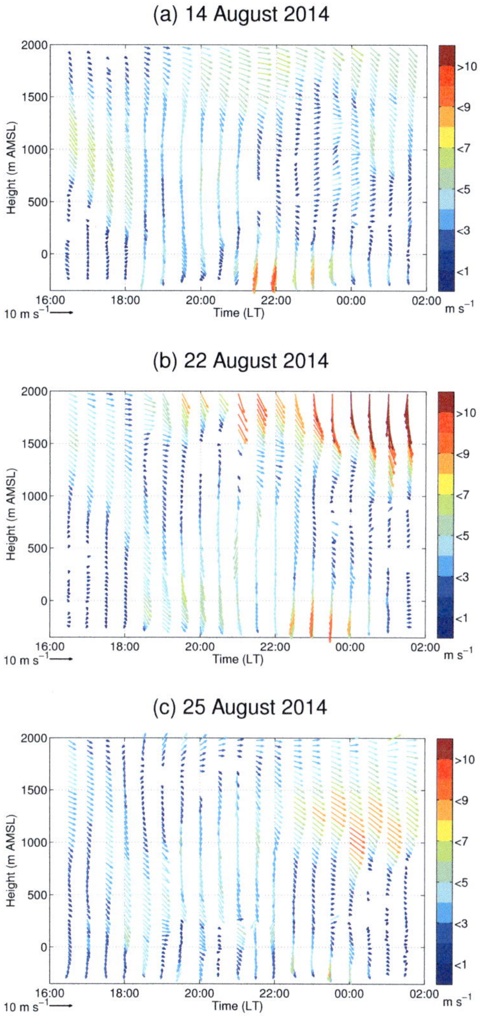

Figure 5.7: Horizontal wind vectors calculated with the VAD algorithm from lidar data at the KITcube site near Masada (Fig. 3.6). For the lowest 500 m Windcube data were used and for the layers above data from the Windtracer. Length and colours of the arrows represent wind velocity and the direction of the arrows indicate the wind direction. 10 min mean profiles for short events lasting only 2 to 3 h are shown.

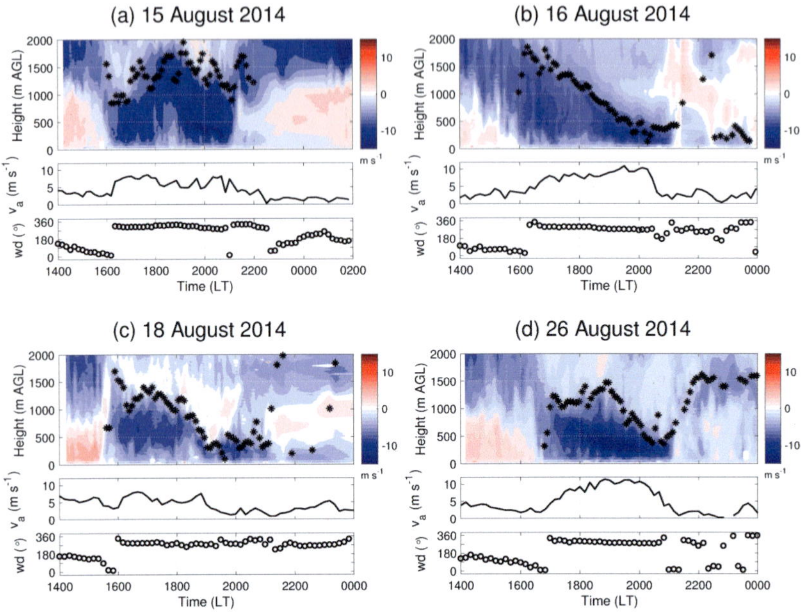

Figure 5.8: Averaged vertical profiles of the horizontal radial wind component derived from lidar RHI scans at 299°(method described in Sec. 4.1). Negative values (blue colours) indicate a wind component from north-west, and positive values (red colours) indicate a wind component from south-east. Black stars indicate the detected height of the layer characterised by severe westerly flow. Below the cross sections wind direction and wind speed measurements in 40 m height measured with the small lidar (Windcube) are shown. Data are shown for four events with a duration of about 4 - 5:30 h.

5.1 to 7.5 m s^{-1}, and mean maximum wind speed varied from 7.5 m s^{-1} on 18 August up to 10.8 m s^{-1} on 26 August (Tab. 5.1). These stronger events did also penetrate further into the valley. The DWF was observed on all of the days at the lidar location with wind speeds in 40 m AGL of up to 8 m s^{-1} on 15 and 18 August and 11 m s^{-1} on 16 and 26 August (Fig. 5.8). Overall, the DWF penetrated at least 13.5 km into the valley with wind speeds up to 10 m s^{-1} (Fig. 5.9), but it could not be determined if it reached the opposite valley sidewalls, as the lidar measurements were limited to a radius of 10 km.

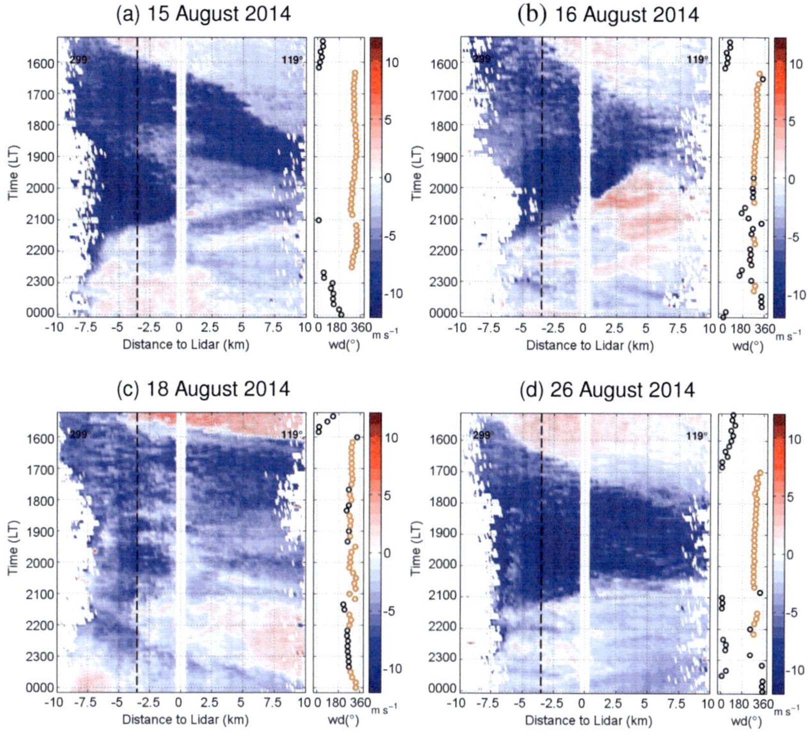

Figure 5.9: Hovmoeller diagrams of the near-surface radial wind velocity derived from lidar RHI scans at an azimuth angle of 119° and 1.6° elevation, and an azimuth angle of 299° and 7° elevation. Negative values (blue colours) indicate north-west wind and positive values (red colours) wind from south-east. The black dashed line indicates the end of the slope and the beginning of the valley floor. Additionally, wind direction measured with the small lidar system in 40 m AGL is shown. Orange points highlight wind directions between 270 and 330°, indicating that the DWF reached the measurement location. Data are shown for the four events with a duration of 4 - 5:30 h.

The DWF stopped around 21:00 - 22:00 LT on all four days even though no along-valley northerly flow was observed afterwards (Fig. 5.4 b, 5.10).

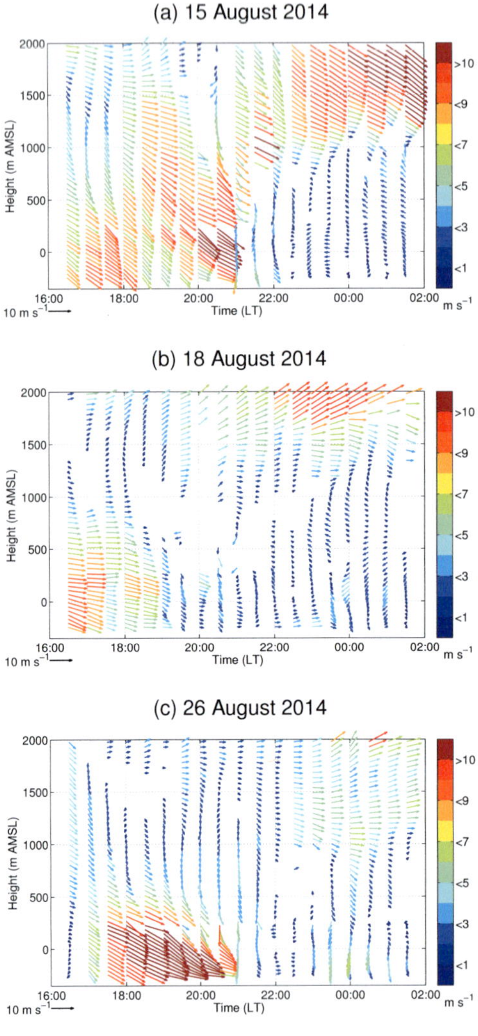

Figure 5.10: Horizontal wind vectors calculated with the VAD algorithm from lidar data at the KITcube site near Masada (Fig. 3.6). For the lowest 500 m Windcube data were used and for the layers above data from the Windtracer. Length and colours of the arrows represent wind velocity and the direction of the arrows indicate the wind direction. 10 min mean profiles for medium events lasting 4 to 5:30 h are shown.

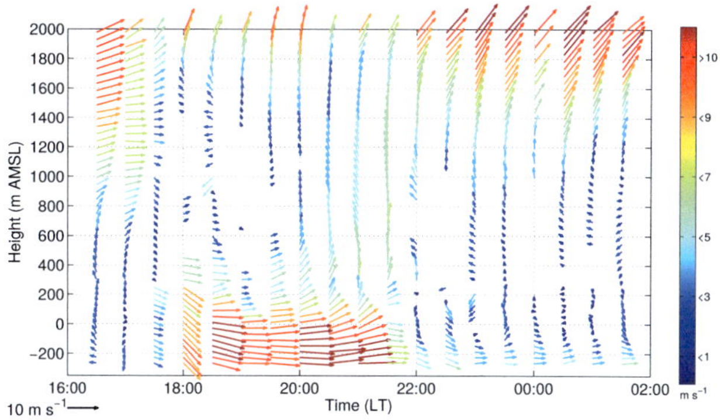

Figure 5.11: Horizontal wind vectors calculated with the VAD algorithm from lidar data at the KITcube site near Masada (Fig. 3.6). For the lowest 500 m Windcube data were used and for the layers above data from the Windtracer. Length and colours of the arrows represent wind velocity and the direction of the arrows indicate the wind direction. 10 min mean profiles for 28 August 2014 are shown.

5.1.3 DWFs of Long Duration

The last two observed events on 28 and 31 August were events lasting over 11 h, which can be associated with pattern P3b. On both days, the DWF started at 18:00 LT and a very shallow layer developed with a mean layer height of 386 m AGL on 28 August and 251 m AGL on 31 August (Tab. 5.1). The mean wind speed was 5.4 m s^{-1} and 4 m s^{-1}, respectively, and mean maximum wind speed reached 6.1 m s^{-1} on 28 August and 8.3 m s^{-1} on 31 August at a height of only 201 m and 124 m AGL, respectively. On 28 August data showed a partitioning of the event into two phases. First, a strong DWF with velocities of 10 m s^{-1} persisted for about 4 h and then transformed into a shallow layer with wind speed around 5 m s^{-1} (Fig. 5.11, 5.12a). Also in the Hovmoeller diagram similarities with the DWF events of medium duration can be seen. For the first 4 h, until 21:30 LT, high wind speeds and a penetration depth of over 13 km were observed (Fig. 5.13). Later, the DWF reached only to the lidar location and the wind velocities were lower (Fig. 5.12). The observations suggest that on

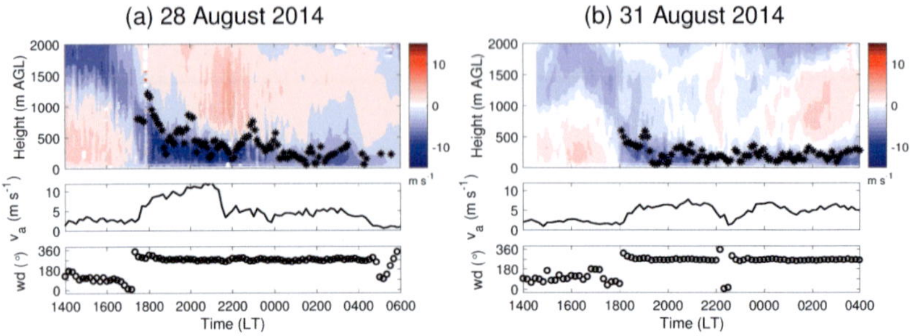

Figure 5.12: Averaged vertical profiles of the horizontal radial wind component derived from lidar RHI scans at 299°(method described in Sec. 4.1). Negative values (blue colours) indicate a wind component from north-west, and positive values (red colours) indicate a wind component from south-east. Black stars indicate the detected height of the layer characterised by severe westerly flow. Below the cross sections wind direction and wind speed measurements in 40 m height measured with the small lidar (Windcube) are shown. Data are shown for the two events with a duration of over 11 h.

28 August a medium DWF event with strong wind speeds transformed into a shallow DWF lasting for the rest of the night.

As can be seen, the systems are rather diverse but what they all have in common are high wind velocities near the ground. They differ in their penetration distance into the valley and in their duration but they all certainly influence the valley atmosphere. To explain the driving forces and the influence of the DWFs on the valley atmosphere, additional measurements, such as vertical profiles of temperature and humidity, are necessary. Therefore, one of the summer IOPs, the 16 August 2014, was chosen as a case study to analyse the driving processes behind the DWF and the effect on the valley atmosphere in Sec. 5.2.

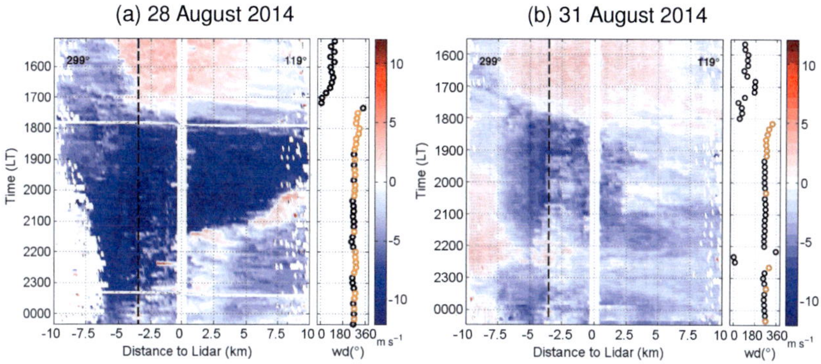

Figure 5.13: Hovmoeller diagrams of the near-surface radial wind velocity derived from lidar RHI scans at an azimuth angle of 119° and 1.6° elevation, and an azimuth angle of 299° and 7° elevation. Negative values (blue colours) indicate north-west wind and positive values (red colours) wind from south-east. The black dashed line indicates the end of the slope and the beginning of the valley floor. Additionally, wind direction measured with the small lidar system in 40 m AGL is shown. Orange points highlight wind directions between 270 and 330°, indicating that the DWF reached the measurement location. Data are shown for the two events with a duration of over 11 h.

5.1.4 Northerly Along-Valley Flow

The last pattern (P4) is a persistent nocturnal north-easterly along-valley flow, which was observed in spring and summer on about 30 to 40% of the nights and in autumn on 10% of the nights. In winter this pattern was not observed. When analysing the windroses of the nighttime data from EBS1 (Fig. 5.14) it can be seen that this northerly flow was connected to wind speeds up to 8 m s^{-1} at the surface, and that it was very dominant in spring and summer. Other wind directions occurring in spring were very weak southerly winds. In summer a secondary westerly maximum can be seen at night. This is caused by the aforementioned long lasting DWFs P3b. For a more detailed analysis of the characteristics of the along-valley flow, lidar data from the summer SOP were used. On 17 days in August 2014 an along-valley flow was observed. Lidar data were available for 10 of those days. Along-valley RHI scans at an azimuth

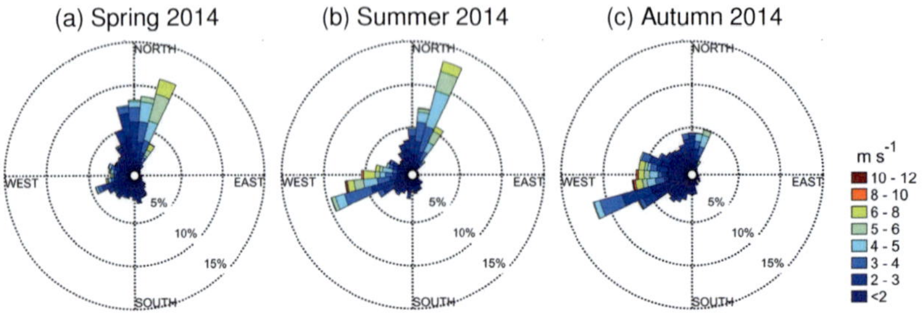

Figure 5.14: Nighttime wind conditions at EBS1 between 20:30 and 6:30 LT.

Table 5.2: Characteristics of the along-valley flows derived from RHI scans at 15° azimuth. Presented are mean radial wind speed of the layer, mean height of the along-valley flow, mean maximum radial wind speed, and mean height of the wind speed maximum. The mean wind speed was first calculated for each time step and then averaged over the whole event. The height, maximum wind speed, and height of the wind speed maximum were determined for each time step from averaged wind speed profiles 2.5 km north of the lidar and then also averaged over the whole duration of the event.

Date	Mean wind speed ($m\,s^{-1}$)	Mean layer height (m AGL)	Mean maximum wind speed ($m\,s^{-1}$)	Mean height of maximum wind speed (m AGL)	Duration
13.08.2014	-4.8	478.0	-7.8	217.0	6:11
14.08.2014	-4.0	571.5	-7.1	200.8	4:54
19.08.2014	-4.0	647.9	-6.6	176.8	8:20
21.08.2014	-6.4	691.8	-9.4	233.5	8:20
23.08.2014	-4.5	783.3	-7.6	451.5	4:01
24.08.2014	-3.0	523.4	-5.4	268.2	5:24
25.08.2014	-3.3	649.7	-5.5	317.2	3:49
26.08.2014	-3.2	596.1	-5.1	302.0	5:30
29.08.2014	-4.6	768.2	-7.8	223.9	7:27
30.08.2014	-4.0	624.5	-7.2	209.8	8:44

angle of 15° were used to analyse the layer height and the speed of the flow using the method described in Sec. 4.1. The flow characteristics were derived from each individual RHI scan and then averaged over the whole flow event for each day (Tab. 5.2). The layer characteristics derived from the RHI scan at an azimuth angle of 353° showed comparable results, and are therefore not shown. The duration of the along-valley flows varied from 4 h up to 9 h, but the onset was always between 21:00 and 22:00 LT. The mean wind speed of the along-valley flow varied between 3 and 4.8 m s^{-1} on nine of the events, only on 21 August the event was stronger with 6.4 m s^{-1} and a mean maximum wind speed of 9.4 m s^{-1} (Tab. 5.2). The mean maximum wind speeds on the other days varied between 5.4 and 7.8 m s^{-1}. The layer height varied between 478 and 780 m AGL and the maximum wind speed was observed near the ground at around 200 m AGL in seven of the cases, at about 300 m AGL on 25 and 26 August and at 450 m AGL on 23 August. No direct correlation between the wind speed and the duration of the events or between the wind speeds and the layer height was found, but a comparison of the onset of the along-valley flow with the along-valley potential temperature gradient resulted in a good match. The potential temperature measured at EBS1 was compared to the potential temperature in Sedom, which is 45 km south of EBS1. In eight cases the along-valley flow started when the temperature difference between the two stations turned negative (Fig. 5.15 a-c and f-j), meaning that the air in the northern part of the valley was colder than the air in the southern part. On some of the nights it was observed that the temperature difference along the valley axis was reduced over the course of the night (or even got positive) causing the northerly along-valley wind to cease after a couple of hours (Fig. 5.15 a,f,g,h,i) . The two cases which could not directly be linked to the along-valley temperature gradient were most likely influenced by large-scale pressure differences. One of the cases was on 21 August, which was already mentioned before, as it had unusually high wind speeds compared to the other cases. Even though the temperature difference between the stations was close to zero (Fig. 5.15 d), a very strong northerly flow was measured in the valley for over 8 h, probably caused by a large-scale pressure gradient. The other case on 23 August was most likely also influenced by the large-scale conditions. On this day a synoptic driven northerly flow at crest height prevailed over the course of the day and was penetrating

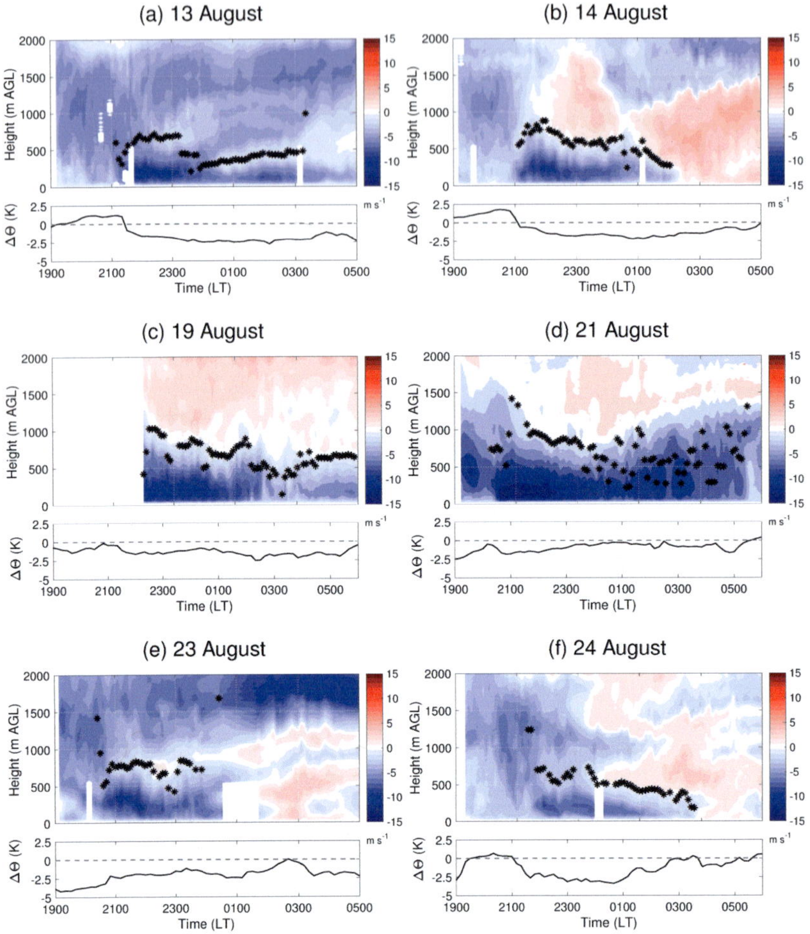

Figure 5.15: Averaged vertical profiles of the horizontal radial wind component derived from lidar RHI scans at 15° azimuth (method described in Sec. 4.1). Negative values (blue colours) indicate a wind component from north, and positive values (red colours) indicate a wind component from south. Black stars indicate the detected height of the northerly along-valley flow. Below the cross sections the potential temperature difference between EBS1 and Sedom (45 km further south) is shown.

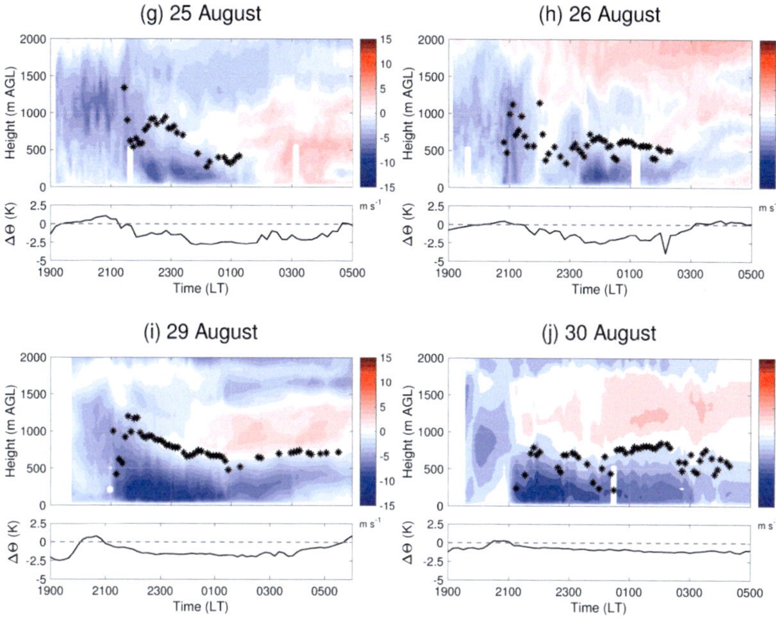

Figure 5.15: (Cont.) Averaged vertical profiles of the horizontal radial wind component derived from lidar RHI scans at 15° azimuth (method described in Sec. 4.1). Negative values (blue colours) indicate a wind component from north, and positive values (red colours) indicate a wind component from south. Black stars indicate the detected flowing layer height. Below the cross sections the potential temperature difference between EBS1 and Sedom (45 km further south) is shown.

into the valley in the evening, most likely caused by a downward momentum transport. It should also be mentioned that a horizontal west-east temperature gradient, and thus a thermal wind parallel to the isotherms, could serve as an additional forcing of the northerly along-valley flow, but this could not finally be established using the measurements. For a further analysis of this along-valley wind systems a detailed analysis of the pressure perturbation along the valley axis and also of the large-scale pressure gradients would be necessary. Unfortunately, no pressure measurements are available for the region north of the Dead Sea. Therefore, COSMO-EU analysis data were used to further investigate the along-valley pressure gradients, but due to the coarse model resolution (7 km) the small pressure perturbations along the valley axis were not resolved and the analysis of the model data did not provide further insights into the development of the along-valley winds.

5.2 Case Study of 16 August 2014

The general aspects of the wind systems were described in Sec. 5.1. Now, a case study is used to further analyse the typical diurnal evolution of the valley ABL connected with a strong DWF in the evening of 16 August 2014. As already seen in Sec. 5.1, the DWF started at around 16:00 LT and had a mean maximum wind velocity of $8.8\,\mathrm{m\,s^{-1}}$. It lasted for about 4 to 5 h and was one of the stronger DWFs which were observed in August 2014. To fully understand the driving processes behind the event, not only the atmospheric conditions in the valley, but also those upstream are important to consider.
The synoptic situation on 16 August represents the typical Israeli summer conditions (Sec. 3.1). A Persian Trough extended from the Persian Gulf over Iraq and Syria bending north-west towards the Mediterranean Sea and Greece (Fig. 5.16). The trough caused a persistent north-westerly flow in the lowest 1000 m AMSL and as the intensity of the trough strengthened in the course of the day near-surface westerly flow over the Mediterranean Sea increased to about 4 - $5\,\mathrm{m\,s^{-1}}$ in the afternoon.

5.2.1 Upstream Conditions

For the analysis of the atmospheric conditions upstream of the valley, only near-surface measurements from stations of the IMS and one vertical profile of the 14:00 LT Bet Dagan (near Tel Aviv) radiosonde were available. To fully understand the development of the DWF in the valley the four-dimensional conditions upstream of the Judean Mountains are important. Therefore, COSMO-EU analysis data, provided by the German Weather Service (DWD), were used.

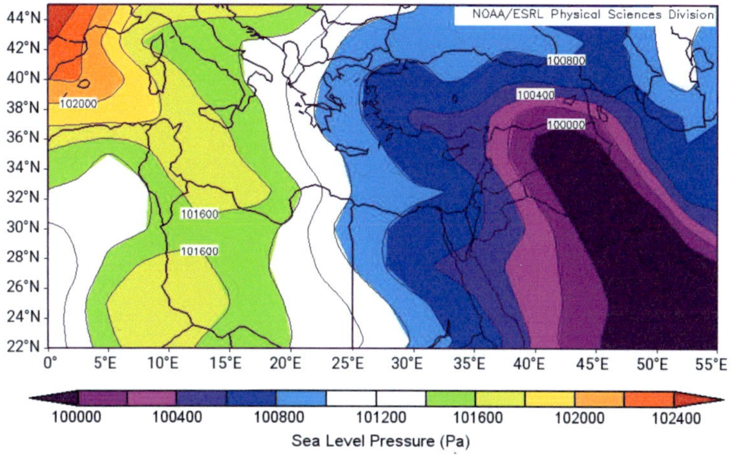

Figure 5.16: Mean sea level pressure (MSLP) in Pa for 16 August 2014 14 LT. Image provided by the NOAA/ESRL Physical Sciences Division, Boulder Colorado (http://www.esrl.noaa.gov/psd/).

The COSMO-EU model is operationally run by the DWD and covers whole Europe with a horizontal resolution of 7 km. Data were available every 3 h and comparisons with near-surface measurements and with the aforementioned vertical profiles of temperature, humidity, and wind showed a fair agreement. In Fig. 5.17 a comparison of the model data with measurements in Bet Dagan and Jerusalem is shown. For Bet Dagan the potential temperature was underestimated by the model by about 1.5 K during the day, but wind direction and wind speed were well represented in the model (Fig. 5.17 a). For Jerusalem, potential temperature was simulated well during the night, but the model did not fully capture the potential temperature increase in the morning. The difference between the model and the measurements was again approximately 1.5 K. The simulated maximum potential temperature at 14:00 LT was well captured again, but afterwards the model temperature decreased too early and was approximately 1.5 h ahead of the measurements (Fig. 5.17 b). The wind direction was well represented in the model for Jerusalem, but the wind speed differed from

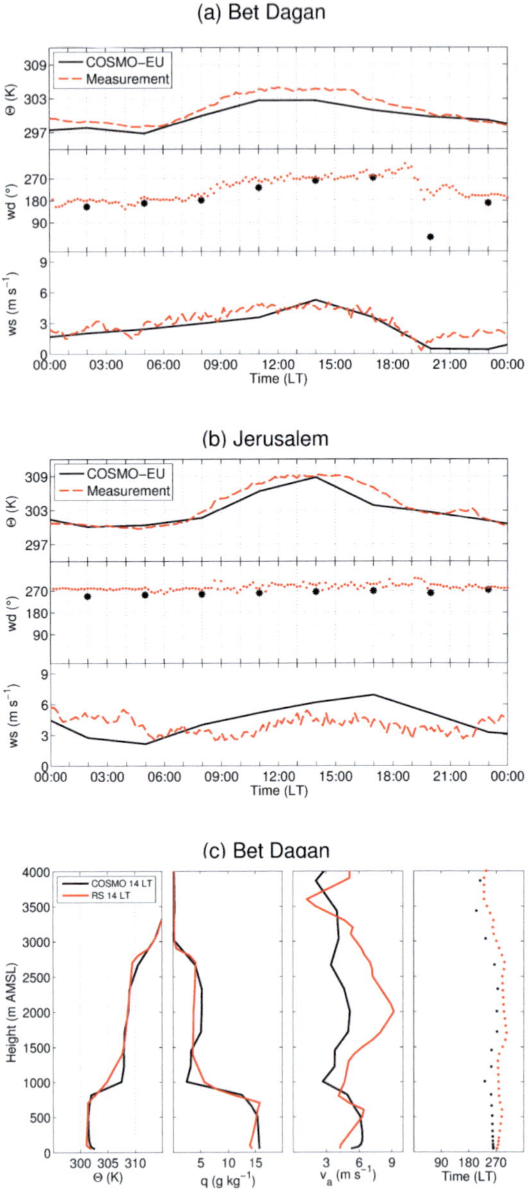
(a) Bet Dagan
COSMO-EU
Measurement
Θ (K)
309
303
297
wd (°)
270
180
90
ws (m s⁻¹)
9
6
3
0
00:00
03:00
06:00
09:00
12:00
15:00
18:00
21:00
00:00
Time (LT)
(b) Jerusalem
COSMO-EU
Measurement
Θ (K)
wd (°)
ws (m s⁻¹)
Time (LT)
(c) Bet Dagan
COSMO 14 LT
RS 14 LT
Height (m AMSL)
4000
3500
3000
2500
2000
1500
1000
500
0
300 305 310
Θ (K)
5 10 15
q (g kg⁻¹)
3 6 9
v_a (m s⁻¹)
90 180 270
Time (LT)

the measurements with an underestimation of nighttime wind velocities and an overestimation of the daytime values. This could either be caused by the limited horizontal model resolution of approximately 7 km, resulting in a limited resolution of the complex terrain and therefore leading to differences in the wind field, or it is caused by the location of the station in the centre of Jerusalem as cities also modify the wind field.

The vertical profiles of potential temperature, specific humidity, and horizontal wind speed and direction were evaluated using radiosonde data from the IMS. The 14:00 LT vertical profile of the potential temperature measured at Bet Dagan shows a good agreement with the modelled profile, considering the CBL inversion which was about 820 m AMSL at 301 K. The inversion strength at the top of the CBL was also well captured with 7 K, only the depth of the temperature inversion was different. However, as only measurements from the main pressure levels and significant points of the vertical profiles are normally provided by IMS no measurement data were available between 900 and 1400 m causing a smooth profile of the potential temperature in this height interval and thus could be responsible for the difference in temperature inversion depth between model and measurements. The modelled specific humidity profile showed small differences to the measured one. First, near-surface humidity was slightly overestimated in the model and between 1500 and 2700 m the model did also have higher values than actually measured. But again, the gradient from the CBL to the free atmosphere was well captured. Wind speed was overestimated near the ground and considerably underestimated in the free atmosphere by the model and the wind direction differed by about 20°. In the model, wind was

Figure 5.17 ***(preceding page)*****:** Potential temperature, wind direction, and wind speed measurements from (a) Bet Dagan and (b) Jerusalem together with COSMO-EU analysis data from the closest grid point for 16 August 2014. Vertical profiles of potential temperature, specific humidity, wind speed, and wind direction, measured with the radiosonde released in Bet Dagan by IMS on 16 August at 14 LT and the respective vertical profiles from COSMO-EU are shown in (c).

coming from west-south-west (≈255°) in the CBL and from west in the free atmosphere, whereas the radiosonde showed a wind direction of 290° in the CBL and a north-westerly flow (≈300°) in the free atmosphere. Even though there are differences between the model and the measurements, the model data represents the conditions upstream of the valley fairly well, especially regarding the CBL height and temperature and will therefore be used to describe the upstream conditions which led to the DWF. The conditions upstream will be further analysed using vertical model profiles of potential temperature, humidity, and wind at two locations. The first point is located north-west of the SOP measurement location directly at the ridge at a height of 687 m AMSL (Fig. 3.1, Ridge), and the second point is near the Mediterranean Sea approximately 13 km away from the shoreline at 62 m AMSL (Fig. 3.1, Upstream). The vertical profiles at those two points are shown in Fig. 5.18. Model data shows that after sunrise the land surface started to heat up and convection slowly eroded the nocturnal boundary layer inversion upstream and at the ridge. Through the vertical mixing a neutral convective boundary layer (CBL) developed (Fig. 5.18). Further heating of the near-surface air mass and strengthening of convective motions warmed the whole boundary layer and the CBL height increased to 965 m AGL over the coastal plains and to 735 m AGL over the ridge at 11:00 LT (Fig. 5.18). The increase of the layer thickness, together with low evaporation rates, caused the specific humidity to decrease. As the large-scale flow was from west with rather strong wind velocities of about 5 to 6 $\mathrm{m\,s^{-1}}$ the evolution of a well defined Mediterranean sea breeze front could not be observed. Neither a temperature decrease nor a humidity increase, the typical characteristics associated with a sea breeze front penetrating inland, were visible over land before 11:00 LT (Fig. 5.19 a). At 14:00 LT wind speed over the sea strengthened and a cooler air mass was advected towards the coast. The stronger onshore wind advected the stratified moist and cool air inland over the coastal plains which hindered the normal evolution of the CBL (Fig. 5.19 b). Potential temperature in the CBL over the plains stagnated at around 302 K and the CBL height decreased to 620 m AGL (Fig. 5.18 a). Over the slopes and at the ridge the evolution of the

CBL was not disturbed by the cooler maritime air mass. Further heating at the ground led to the diurnal maximum of potential temperature in the CBL, which was 308 K at the ridge and weakened the inversion strength at the top of the CBL, which was at 910 m AGL (Fig. 5.18 b). Later, with decreasing solar radiation, convection weakened and potential temperature decreased. The mean potential temperature in the CBL upstream was 300.6 K, and at ridge height it was 304.3 K at 17:00 LT, which would actually be 18:30 LT as there was a time difference of about 1.5 h in the evening were the model temperature decreased too early. At the same time, the boundary layer inversion near the coast and at the ridge strengthened to 4.1 K and 5.0 K, respectively, and the mean wind speed in the boundary layer strengthened to 5.1 $m\,s^{-1}$ near the coast and 7.7 $m\,s^{-1}$ at the ridge (Fig. 5.18). The CBL height decreased considerably to 360 m AGL over the plains and 330 m AGL over the mountain ridge (Fig. 5.18). This results in a Froude number (Eq. 2.9) of 0.73 over the plains and 1.06 at the ridge indicating a transition of a subcritical flow towards a critical stage at ridge height. Fig. 5.19 (c) shows that the moist and cold maritime air mass penetrated inland over the coastal plains but stopped at the foot of the mountains. With further radiative cooling of the surface the boundary layer got stratified and the upstream air mass got blocked by the mountain ridge as indicated by a non-dimensional mountain height (Eq. 2.3) of $\hat{H} \approx 2.7$ at 20:00 LT in the model. The conditions upstream of the valley already show potential for a strong DWF event in the evening. The advection of the maritime air mass over the plains resulted in an inhibited boundary layer evolution near the coast in the afternoon, whereas over the ridge the conditions were undisturbed. A stronger decrease of the CBL height over the ridge compared to the boundary layer height decrease near the coast then resulted in the transition of a subcritical flow over the plains ($Fr = 0.73$) to a critical stage over the ridge ($Fr = 1.06$), which indicates a supercritical flow into the valley. The conditions, which enable such a supercritical flow, changed in the night, when further radiative cooling finally led to a blocking of the upstream air mass at the mountains ($\hat{H} \approx 2.7$) and the air was not longer able to flow over the ridge into the valley.

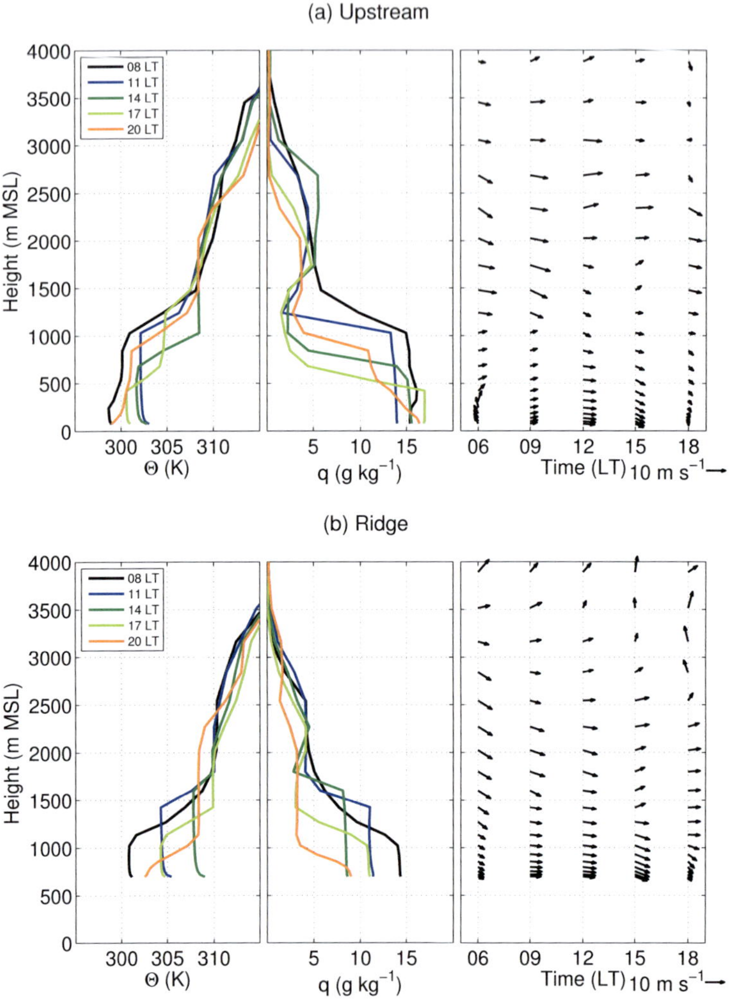

Figure 5.18: Vertical profiles of potential temperature, specific humidity, and wind at the location upstream near the coast (a) and at the ridge (b) from COSMO-EU analysis data for the 16 August 2014.

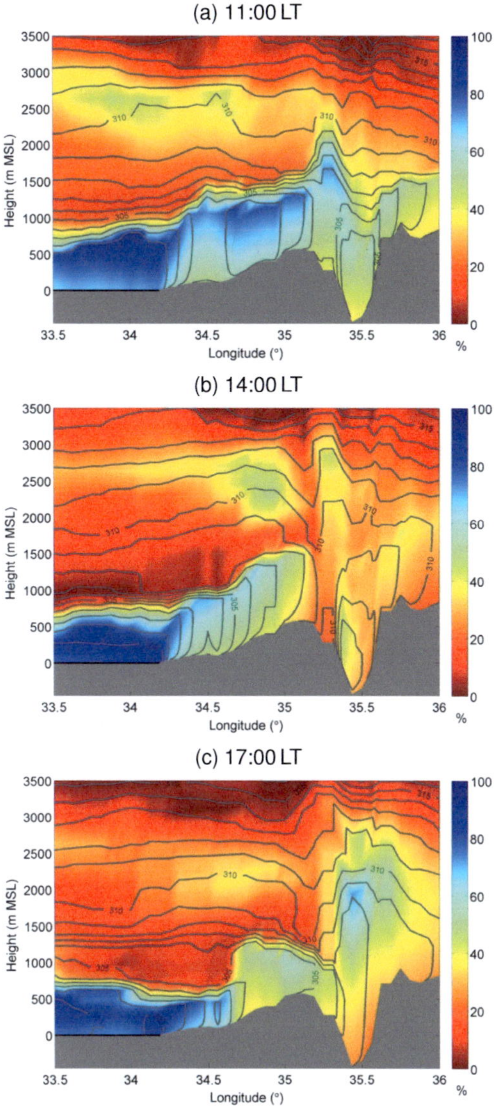

Figure 5.19: West-east cross section of relative humidity (coloured) and potential temperature (isolines) at the latitude of Masada from COSMO-EU analysis data for 16 August 2014.

5.2.2 Valley Conditions

The description of the daytime atmospheric boundary layer (ABL) conditions in the valley can be divided into three time periods: the time prior to the DWF between 07:00 LT and 15:00 LT, the development of the DWF in the afternoon between 15:00 LT and 18:30 LT, and the hydraulic jump in the evening between 18:30 LT and 21:00 LT. Afterwards, the DWF was no longer observed.

CBL Evolution Prior to the DWF

On 16 August upslope winds developed at 07:00 LT. The measurements at the slope (Masada) showed a change in wind direction from westerly to easterly, and the vertical wind velocity turned positive to $w \approx 0.4\,\mathrm{m\,s^{-1}}$ at the same time (Fig. 5.20). The typical lake breeze, already described in Sec. 5.1, reached the KITcube location at 09:00 LT, which can be seen in a strengthening of the 2 m wind speed to about $2\,\mathrm{m\,s^{-1}}$, easterly wind directions, and the attenuation of the diurnal temperature increase, caused by the advection of the cooler air from the water towards the land (Fig. 5.20). The moderate wind speeds caused an evaporation at the lake surface of about 0.05 to $0.07\,\mathrm{mm\,30\,min^{-1}}$, as shown by the data of EBS2, close to the shoreline (Fig. 5.21) and the IWV measured at the KITcube location reached values of about $30\,\mathrm{kg\,m^{-2}}$ (Fig. 5.20). The vertical profiles of potential temperature and wind measured with a radiosonde at 9:00 LT showed that the CBL was up to 900 m AMSL high and that the onshore component of the lake breeze extended to 350 m AMSL with a mean wind speed of $2.4\,\mathrm{m\,s^{-1}}$ (Fig. 5.22 a). The return flow of the lake breeze circulation was in the layer between 500 and 900 m AMSL with a south-westerly wind direction. Above, a strong north-westerly flow prevailed with a wind speed of about $11\,\mathrm{m\,s^{-1}}$. Strong vertical wind shear between the easterly flow in the valley and the strong westerly flow above resulted in a Richardson number (Eq. 2.29) of less than the critical value of $Ri_c = 0.25$ in the layer between 350 and 900 m AMSL at 9:00 LT. Consequently, the induced turbulence led to a downward mixing of warmer air from above in the time between 9:00 and 11:00 LT and resulted in an

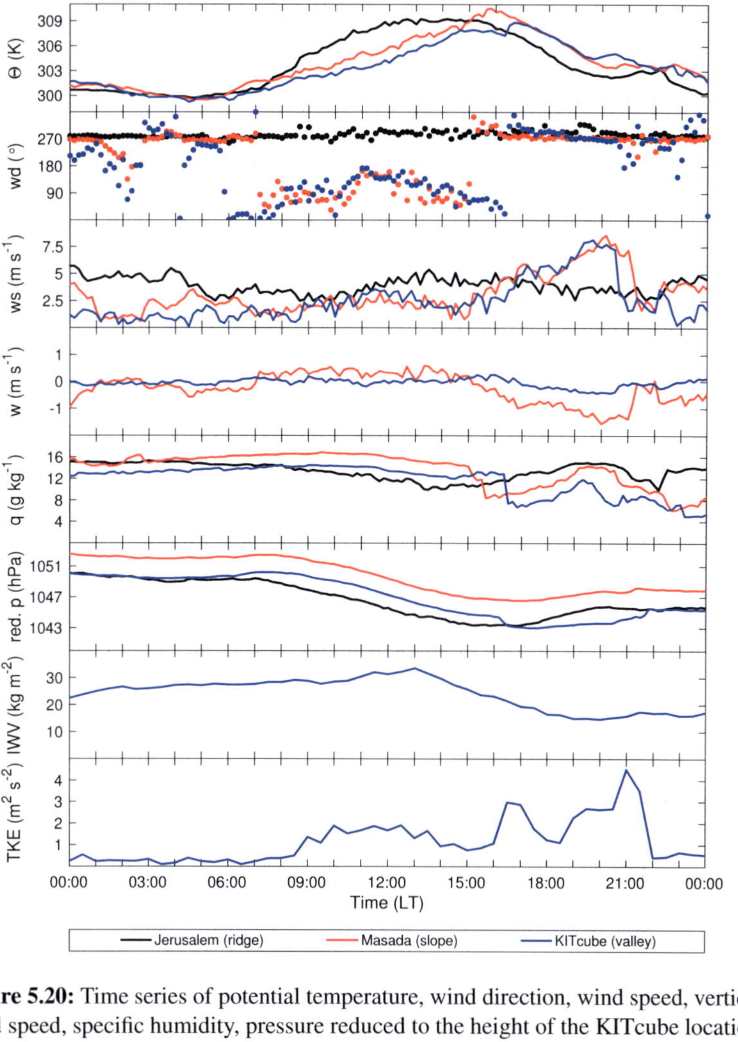

Figure 5.20: Time series of potential temperature, wind direction, wind speed, vertical wind speed, specific humidity, pressure reduced to the height of the KITcube location, IWV, and turbulent kinetic energy (TKE) for Jerusalem, Masada, and the KITcube location. Data from Jerusalem represent the conditions at ridge height, Masada data show the conditions at the slopes of the Judean Mountains, and data from KITcube location show the conditions in the valley on 16 August 2014.

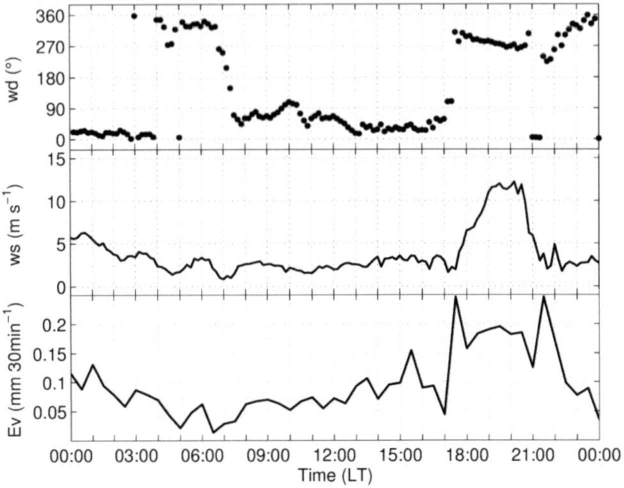

Figure 5.21: Time series of 30 min mean values of wind direction, wind speed, and evaporation at EBS2 for 16 August 2014.

additional heating of the CBL in this layer (Fig. 5.22 b). This stabilised the previously near neutral boundary layer in the upper part and an inversion of about $\Delta\Theta \approx 2$ K established around 550 m AMSL. At 1200 m AMSL a secondary weak inversion represented the former CBL top until 15:00 LT. Evaporation increased and varied between 0.07 and 0.1 mm 30 min^{-1} between 11:00 and 15:00 LT (Fig. 5.21) and through the increased moisture input to the atmosphere the IWV also increased to values of about 35 kg m^{-2} (Fig. 5.20). The mass concentration of small particles with a diameter smaller than 0.35 µm decreased from 4 µg m^{-3} at 11:00 LT until it reached about 2.5 µg m^{-3} at 12:00 LT and was then constant until 16:30 LT (not shown).

Development of the DWF

At 15:00 LT the potential temperature at the valley floor was 308 K. At the slopes it was even 309 K and still increasing (Fig. 5.20, 5.22 a). However, at the ridge (Jerusalem) the temperature already started to decrease and also reached 309 K.

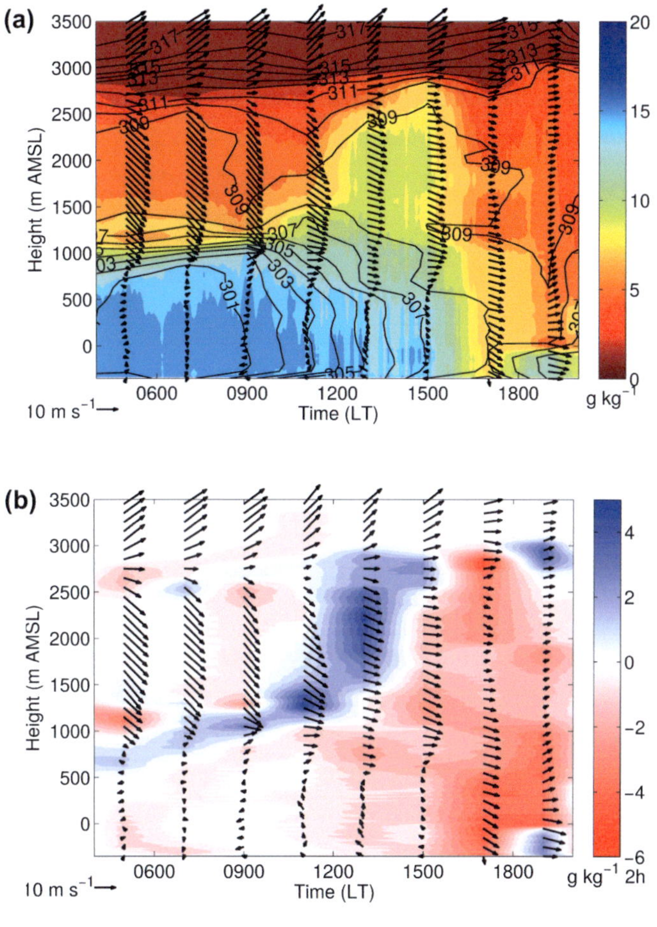

Figure 5.22: Potential temperature (isolines), specific humidity (coloured) and horizontal wind, measured with radiosondes at the KITcube location (a). Respective drying and moistening of the atmosphere between two radiosonde launches calculated from the vertical humidity profiles, and vertical profile of horizontal wind (b). Data are shown for 16 August 2014.

Around the same time the wind direction at the slopes suddenly changed to west, horizontal wind speed increased, specific humidity dropped by 5.8 g kg^{-1}, and vertical velocity became negative, indicating the intrusion of a dry downslope flow with accelerating wind speed into the valley (Fig. 5.20). About 1.5 h later, at 16:30 LT, the DWF finally reached the valley floor. Wind direction changed to 270°, wind speed increased considerably to about 5.5 m s^{-1}, potential temperature increased to 309 K, specific humidity dropped by 5.5 g kg^{-1}, and TKE increased from 1 to 3 m^2 s^{-2} at the location of KITcube (Fig. 5.20). Through the acceleration of wind speed the dynamic air pressure increased, causing a reduction of the measured static air pressure of 2 hPa (Fig. 5.20). The intruding air mass was considerably drier than the valley air mass and had less small aerosol particles. The particle mass of small particles between 0.23 and 0.35μm decreased to 1.2μg m^{-3}, and as the intruding air mass replaced the air in the valley, the IWV dropped from 33 to 17 kg m^{-2} (Fig. 5.20). About 12 km further to the north the DWF reached the Dead Sea at about 17:30 h, which can be seen at the sudden change in wind direction and the increase in wind speed at EBS2 (Fig. 5.21). Through the increase in wind speed the evaporation from the water surface increased considerably from 0.05 to 0.25 mm 30 min^{-1} (Fig. 5.21). The vertical extent of the intruding air mass was estimated with the algorithm described in Sec. 4.1, using lidar RHI scans along the main flow axis of the DWF (299°). The DWF reached the valley 3.5 km north-west of the lidar at about 16:00 LT, 30 min before it was measured at the KITcube location, with a mean height of about 1750 m AGL and a horizontal radial mean wind velocity of about 6.0 to 7.0 m s^{-1} (Fig. 5.23). The layer height varied between 1400 and 1750 m AGL and the horizontal radial mean layer wind speed was about 6.5 m s^{-1} (Fig. 5.23). The vertical profile of specific humidity and potential temperature measured by a radiosonde launched at 17:00 LT showed a dry and near neutral layer up to 900 m AMSL (1250 m AGL) with the same potential temperature of 308 K as was measured in the layer between 550 and 1100 m AMSL at 15:00 LT, and specific humidity was about 9 g kg^{-1} (Fig. 5.22). This is another evidence that the air from the flowing layer replaced the air in the valley. From

17:00 until 18:30 LT the downslope layer flow continued in a relatively steady state with a layer height of about 1050 to 1400 m AGL and a horizontal radial mean wind velocity of 6.0 to 6.5 m s^{-1} (Fig. 5.23).

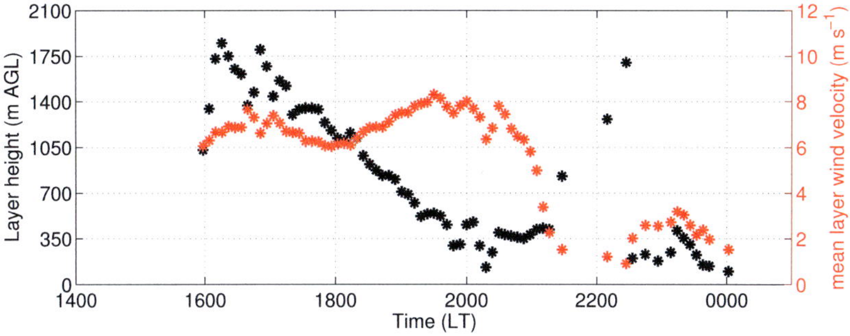

Figure 5.23: Time series of the 10 min running mean of the height and the horizontal radial mean wind velocity of the DWF, calculated from lidar RHI scans on 16 August 2014.

Hydraulic Jump

After 18:30 LT the mean horizontal radial wind velocity of the flow increased to about 8 m s^{-1}, and consequently, the height of the DWF decreased to about 350 m AGL (Fig. 5.23). Also the near-surface measurements showed a further increase of the horizontal wind speed and increasing turbulence (Fig. 5.20). A radiosonde launched at 19:00 LT showed a strong temperature inversion of 3.3 K at 270 m AGL (-80 m AMSL) and the potential temperature in the DWF layer was 305.6 K (Fig. 5.22). The mean horizontal wind speed in the layer measured with the radiosonde was about 9.1 m s^{-1} and slightly higher than the horizontal radial wind speed calculated from the RHI scans (Fig. 5.23). The Froude number of hydraulic flow theory (Eq. 2.9) for this layer is calculated to $Fr \approx 1.7$, indicating a supercritical flow. As the DWF replaced the near-surface air mass in the valley, the IWV remained low and the concentration of small particles remained small as well. Shortly after 19:00 LT, about 4.4 km east of the lidar, a sudden increase of the DWF layer height and a reverse flow near the surface was identified in the RHI scans. It rapidly increased in depth, and at about 19:30 LT the height of the layer was already at about 900 m AMSL (Fig. 5.24), and the air below was quite turbulent. These are indicators for a hydraulic jump with a rotor formation beneath, caused by supercritical flow conditions. The backscatter RHI scan shows that the air mass in the reverse flow has the same aerosol content as the air mass in the DWF layer near the surface, indicated by the same high backscatter values of -7.7 dB. The high aerosol content results from local dust emissions, caused by the high wind velocities and the increased turbulence of the DWF (Fig. 5.20). The air above the DWF has lower and quite uniform backscatter values of -8.25 dB, representing another air mass with another aerosol content. This findings support the assumption of a rotor beneath the hydraulic jump. Through the strong reverse flow of the rotor a convergence line formed in the valley. This can be seen in the PPI scan performed at 0.2° elevation, which shows the wind field at 19:51 LT. North of the lidar, the convergence line is located further to the east close to the shoreline of the Dead Sea, and south east of the lidar it was only 2 km away, which is also supported by the spectrum width data in Fig. 5.24 d. The spectrum width provides a measure of the variability of the mean radial velocity in a measurement bin of the lidar due to wind shear and turbulence. The convergence line can clearly be identified along the increased spectrum width values, showing high wind shear and turbulence along the convergence line. The different location of the convergence line north and south-east of the lidar results most likely from the specific orographic characteristics west of the lidar. North-west of the lidar a wadi opens into the Dead Sea valley. The westerly flow is most likely channelled in the wadi, resulting in accelerated wind speeds, and thus a stronger penetration into the valley. Over the next hour the convergence line moved towards the western mountain ridge and the rotor grew horizontally and vertically. The convergence line passed the KITcube location at 21:00 LT, which can be seen in

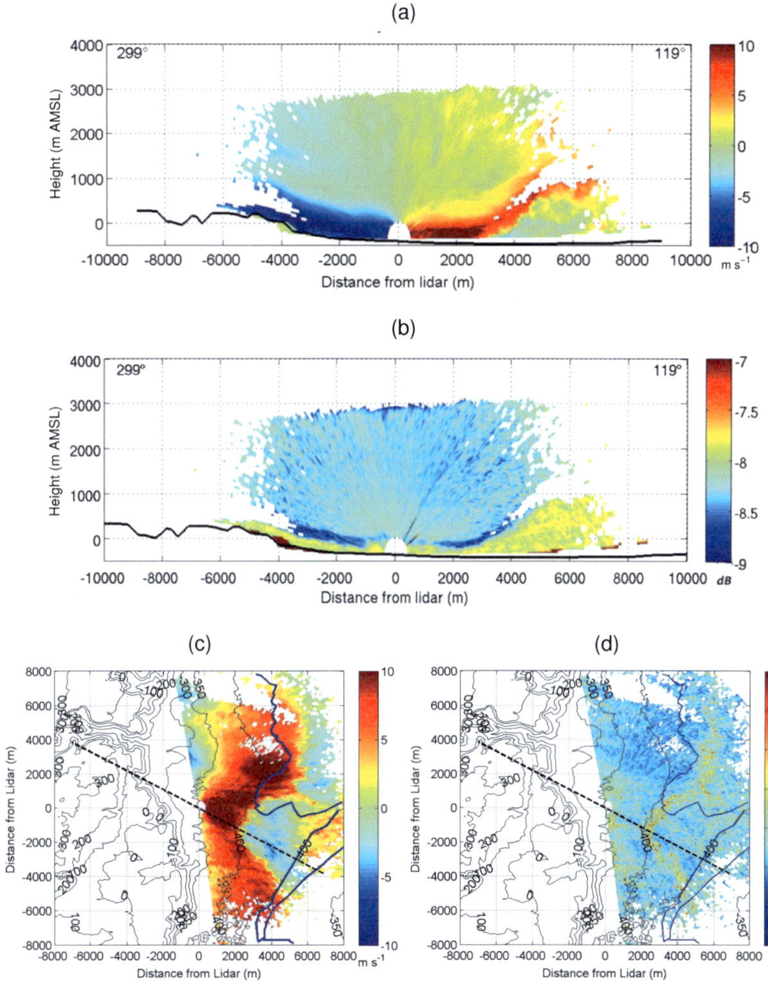

Figure 5.24: RHI scan at an azimuth angle of 119° showing radial velocity (a) and backscatter (b) on 16 August 2014 19:36 LT. PPI scan at an elevation angle of 0.2°. Shown is radial velocity (c) and spectrum width (d) at 19:51 LT. In (a) and (c) blue colours indicate a flow towards the lidar and red colours indicate a flow away from the lidar. The shoreline of the Dead Sea is indicated by a dark blue line in (c) and (d).

maximum TKE values of $4.6\,m^2\,s^{-2}$ (Fig. 5.20). The DWF retreated towards the mountains until it ceased around 21:30 LT (Fig. 5.9, Fig. 5.20).

5.2.3 Conceptual Model

The evolution of the ABL in the valley is influenced by several processes. The evolution of the DWF in the Dead Sea valley depends on the ABL conditions in the valley but also on the upstream conditions west of the mountain ridge. To describe and illustrate the temporal and spatial interactions, a conceptual model was developed (Fig. 5.25). The diagrams show the temporal evolution of the different layers and wind systems and the atmospheric conditions in four stages on 16 August 2014. The boundary layer evolution before the DWF is described in the morning and midday stage, the evolution of the DWF takes place in the afternoon, and the hydraulic jump in the evening. For other days with DWFs the strength and timing of the different processes can vary.

Morning In the morning the ABL in the valley was decoupled from the flow above due to a strong temperature inversion around crest height. Undisturbed by the upstream conditions, upslope winds developed shortly after sunrise, and only 2 hours later a local lake breeze developed (Fig. 5.25). Above the inversion, which decoupled the convective boundary layer in the valley from the large-scale flow, a strong synoptic-driven westerly flow prevailed. As a result of the evolution of the lake breeze in the valley, a strong wind shear between the easterly lake breeze and the strong westerly flow aloft evolved. This caused mechanically produced turbulence at the top of the valley CBL, which was indicated by Richardson numbers below the critical value of 0.25 (striped area in Fig. 5.25). Owing to the turbulence, warmer air was mixed into the upper part of the valley CBL, and a secondary temperature inversion below ridge height evolved. Even though the turbulent mixing decreased the inversion height, an undisturbed valley ABL existed below the inversion.

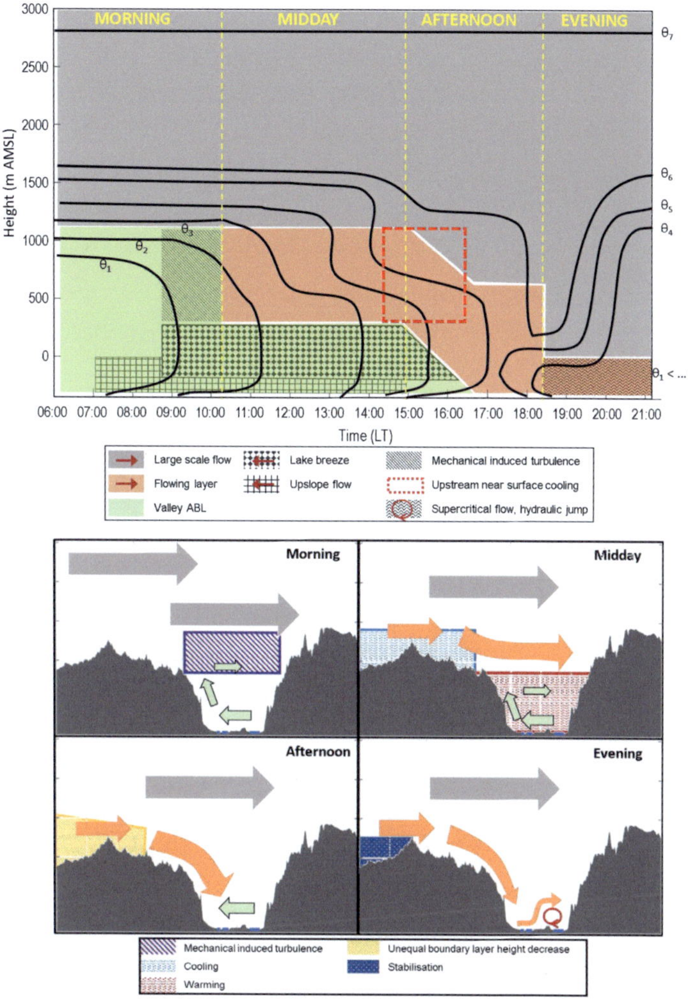

Figure 5.25: Conceptual model describing the different processes governing the Dead Sea valley flow on 16 August 2014. The temporal evolution of the different layers and wind systems is given in the top panel. The wind systems (arrows) and the ongoing processes (shaded areas) leading to the next stage for morning, noon, afternoon, and evening, are shown in the lower panel.

Midday Between the so created temperature inversion below ridge height and the temperature inversion at the CBL top a westerly flow with reduced wind speeds established and formed a flowing layer between the valley CBL and the large-scale strong westerly flow around midday. Below the inversion the valley CBL was further heated, which was mainly caused by the two factors described in Sec. 2.1.3. Firstly, due to the existing inversion in the valley, the lower valley atmosphere was decoupled from the large-scale flow, and radiative heating was concentrated within this reduced air volume in the valley. Secondly, the topographic amplification factor (TAF) also caused a stronger heating of the valley atmosphere.

Afternoon The so induced stronger heating together with the simultaneous cooling outside of the valley and around ridge height finally eroded the inversion below ridge height and a density current formed. The cooler air in the flowing layer formed a relatively deep DWF and while it descended into the valley wind speed accelerated. A change in the upstream conditions triggered the further events. The wind speed increased at the ridge, whereas it decreased over the plains, leading to a stronger decrease of the CBL height over the ridge than over the plains. This already indicated the transition of a subcritical flow regime over the coastal plains to a critical flow regime at ridge height, which was supported by the Froude numbers which were 0.73 over the coastal plains and 1.06 at the ridge at 18:30 LT (model data from 17 LT, considering the time shift of 1.5 h of the model).

Evening The transition of the subcritical flow to the critical state at ridge height finally led to a supercritical flow downstream. This was confirmed by the Froude number of the DWF layer in the valley, which was 1.7, calculated from the radiosonde at 19:00 LT. Downstream in the valley the flow went suddenly from its supercritical to a subcritical state. This transition is also known as a hydraulic jump. It increased the layer height and decreased the wind velocity by generating severe turbulence. Below the hydraulic jump a rotor formed leading to a near-surface convergence line with high turbulence in the valley. Further cooling over the coastal plains finally led to a blocking of the upstream air

mass by the mountains, as indicated by a non-dimensional mountain height of $\hat{H} = 2.7$ at about 21:30 LT, and thus stopped the flow over the mountains and the DWF in the valley.
As can be seen, the preliminary trigger of the strong DWF was the cooling of the air at ridge height and the simultaneously prolonged heating within the valley forming a density current. However, the intensification of the DWF, the hydraulic jump, and the rotor formation were caused by processes upstream of the valley, leading to the transition of the subcritical flow to a supercritical flow. While over the ridge the CBL evolution was undisturbed during the day, the evolution of the CBL over the coastal plains was inhibited by the Etesian flow, which advected cooler maritime air inland. Due to the presence of the different air masses the CBL height decreased stronger over the ridge compared to the coastal plains in the evening, leading to the transformation of a subcritical flow over the coastal plains to a critical state at the ridge and a supercritical flow in the Dead Sea valley.

5.3 Case Study of 5 to 8 November 2014

In Sec. 5.1 it was already discussed that in winter the synoptic conditions have a larger influence on the valley atmosphere and that the wind conditions vary strongly. Still, strong wind events can be observed which are caused by the complex orography, and which have a strong impact on the atmospheric conditions. An strong event was observed in November 2014. Between 6 and 8 November a northerly wind with velocities of up to 10 m s^{-1} during daytime was observed in the valley. In that period relative humidity dropped to 15 %, and daily evaporation from the Dead Sea water surface was 5.12 mm d^{-1}, 6.86 mm d^{-1}, and 4.55 mm d^{-1}, respectively. These evaporation rates were more than three times higher than the median evaporation for November and the highest daily evaporation amounts observed in the investigation period from March 2014 until March 2015. Therefore, this event shall be discussed in detail in this section.

5.3.1 Large-Scale Conditions

The large-scale conditions are analysed using 3 hourly Global Forecast System (GFS) analysis and data from the IMS station in Jerusalem. According to the semi-objective synoptic classification from Alpert et al. (2004) a Red Sea Trough with a Central axis (RSTC) formed on 5 November around 14:00 LT (Fig. 5.26). It intensified over the next 24 h leading to a change of the near-surface air flow outside the valley from west to north on 4 November. This change was also observed in Jerusalem, where the wind direction changed from 270 to 350° (Fig. 5.27). A high-pressure system over Syria north-east of Israel intensified and propagated south leading to a stronger surface pressure gradient, oriented from north-east to south-west over the investigation area and a strong easterly flow over the Dead Sea valley (Fig. 5.26). In Jerusalem wind direction changed from north to east over the course of the day on 6 November and became nearly constant with a wind direction of about 90° to 100°, after 20:00 LT (Fig. 5.27). The surface pressure gradient intensified over the next 63 h until 9 November 5:00 LT and led to strong south-easterly wind over whole Israel, which advected warm and very dry air from the Arabian Desert (Fig. 5.26). Later, the pressure gradient and subsequently near-surface wind velocities weakened, which can also be seen in the measurements in Jerusalem.

5.3.2 Valley Conditions

To investigate the influence of the orography, observations at two stations, one on the slopes, and one on the valley floor are analysed.

Slope Station

For the analysis of the conditions higher up in the valley meteorological data from MET1, which is located at 124 m AMSL on the Jordanian side of the Dead Sea (Fig. 3.1), are used and shown in Fig. 5.27. On 6 November at 00:00 LT, specific humidity started to decrease considerably from 9.3 to approximately 5 $g\,kg^{-1}$. However, the wind field still showed the typical periodic wind shift

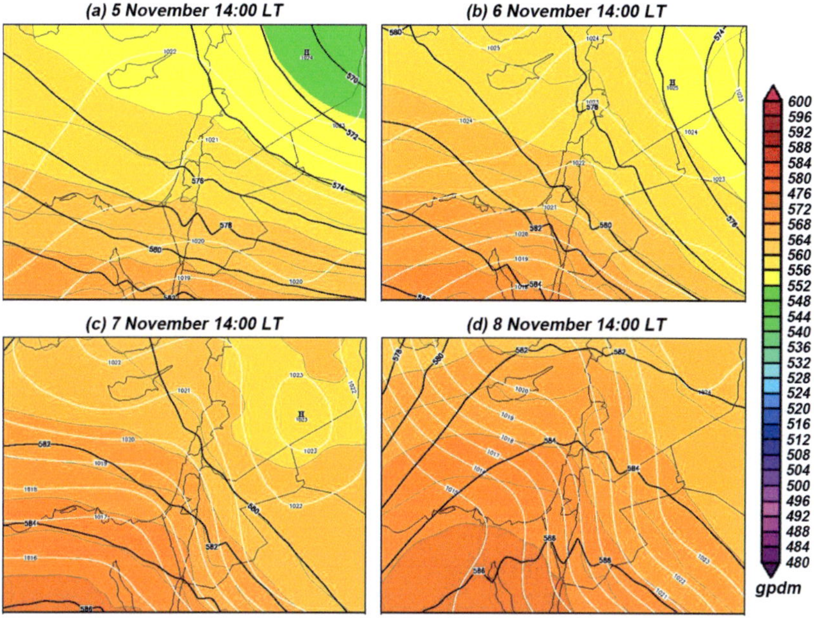

Figure 5.26: Evolution of 500 hPa geopotential (black isolines), surface pressure (white isolines), and relative topography 500-1000 hPa (coloured) from 5 November until 8 November 2014. Data from GFS analysis, ©Wetter3.de

from south-east to north, over the course of the day as well as moderate wind velocities below 6 m s^{-1} (Fig. 5.27). At 18:00 LT on 6 November wind speed increased suddenly to values between 5 and 10 m s^{-1}, wind direction turned from north to south-east and specific humidity dropped further to 3.2 g kg^{-1}. From 6 until 9 November 7:00 LT, the wind direction remained mainly south-easterly with high wind velocity. This strong south-easterly flow was only interrupted for time intervals of about 4 h, from 10:30 until 16:30 LT, on 7 and 8 November (Fig. 5.27). During this time intervals upslope winds caused westerly winds with low velocities, which advected slightly moister air from the valley floor to the station (Fig. 5.27). Between those two time intervals the wind was even

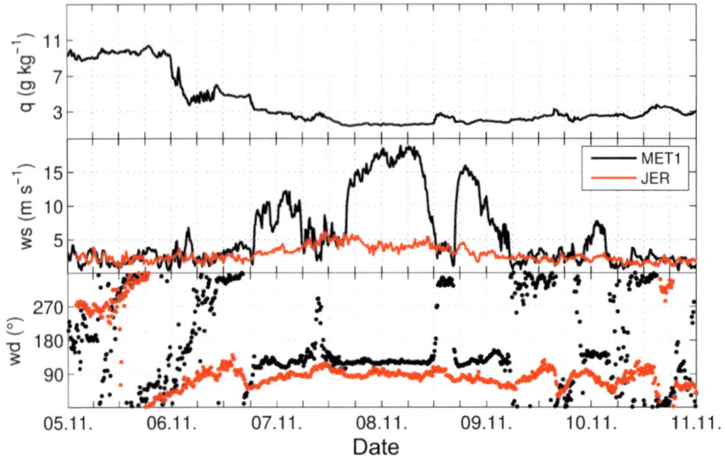

Figure 5.27: Time series of 30 min mean meteorological variables measured at MET1 (black), which is located at 124 m AMSL on the Jordanian side of the Dead Sea, and in Jerusalem (red). From top to bottom: specific humidity, wind speed, and wind direction.

stronger than on 6 November with values between 12 and 19 m s^{-1}, and specific humidity decreased to approximately 1.7 g kg^{-1}.

Valley Station

In Fig. 5.28 the time series of wind speed and other relevant meteorological parameters averaged over 30 min at the location of EBS2 (in the valley) are shown. Additionally, IWV derived from GPS data are shown for the same time and location. Already on 5 November at 13:00 LT wind direction turned north, but averaged wind speed remained moderate, with values between 1.8 and 6 m s^{-1}. During the night wind speed strengthened. At the same time, specific humidity started to decrease and vapour pressure deficit increased to 13 hPa. The IWV also started to decrease, which indicates that not only the near-surface air became drier but also the air in the upper levels. On 6 November at 9:00 LT wind speeds of 9 m s^{-1} were measured. This caused an increase of evaporation from 0.04 to 0.17 mm 30 min^{-1}. The atmospheric conditions remained similar over the next two days. A constant northerly wind, with fluctuating but relative high wind velocities, advected warm and dry air into the Dead Sea valley. The vapour pressure deficit ranged between 11 and 21 hPa, and 30 min mean wind speed even exceeded 10 m s^{-1}. This conditions caused high evaporation rates with maximum values of 0.24 mm 30 min^{-1} on 7 and 8 November at 9:00 LT. The IVW decreased continuously from the initial 30 kg m^{-2} on 5 November to only 4 kg m^{-2} on 7 November. Later, it increased slowly to about 10 kg m^{-2} and remained at this level for the rest of the period. The air mass advected from the north was not only very dry, but had also a very low aerosol content. The AOD showed a decrease from the initial 0.2 on 5 November to 0.06 during the event, and visual inspection of photographs showed a very clear near-surface air mass with no haze. Even though the main wind direction was northerly, diurnal changes of wind conditions were observed. During daytime the wind direction was north-north-east and wind speed was high, and during nighttime the wind speed was lower and the wind direction shifted to north-north-west. The daily maximum of the wind velocity was reached around 9:00 LT, matching the maximum evaporation rates on these consecutive days. After 9:00 LT on 8 November, both wind velocity and evaporation started to decrease, and when the wind turned to west at 16:30 LT, wind velocity was only 0.3 m s^{-1} and the evaporation rate 0.01 mm 30 min^{-1}. At 19:00 LT a northerly flow set in again. As a result, wind turned back to north, wind speed increased and also evaporation increased again. Finally, on 9 November around noon, wind turned to south and wind velocity decreased to lower values between 1 and 3 m s^{-1}. The vapour pressure deficit and evaporation rate decreased respectively (Fig. 5.28).

The analysis of the vertical profiles of horizontal wind speed and direction measured by the lidar systems in the southern part of the valley showed that the pressure-driven forcing reached at least up to about 350 m AGL (Fig. 5.29). Above, the wind turned clockwise, and above ridge height the geostrophic wind direction was reached with wind velocities of about 4 to 7 m s^{-1}.

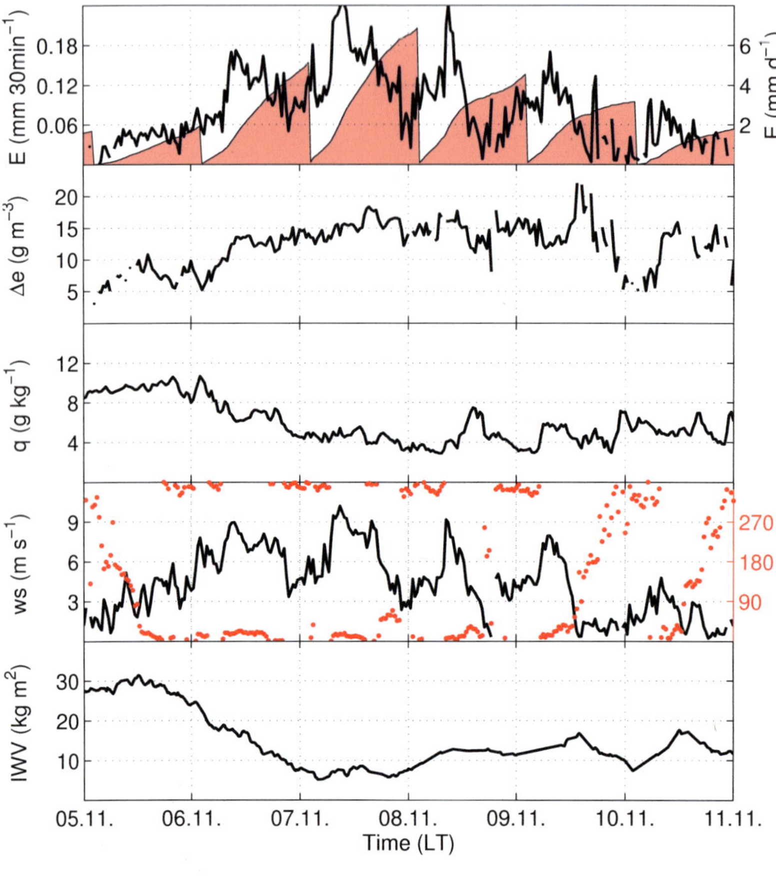

Figure 5.28: Time series of 30 min mean meteorological variables measured at EBS2. From top to bottom: evaporation and accumulated daily evaporation sum (coloured), vapour pressure deficit, specific humidity, wind speed (solid line) and wind direction (red points), and integrated water vapour measured with the GPS.

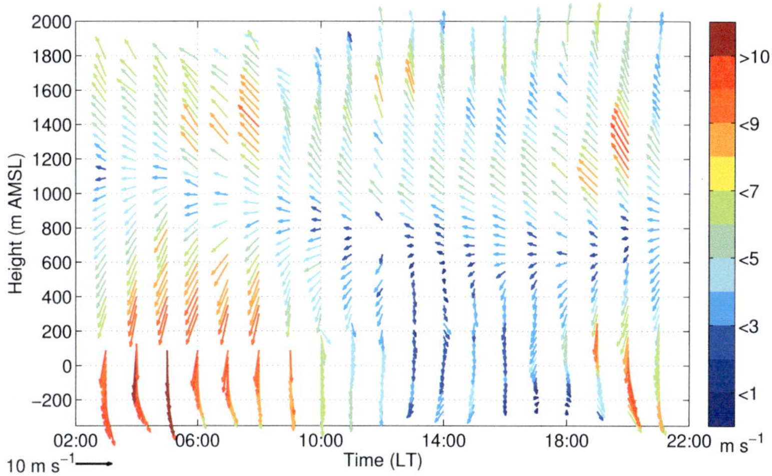

Figure 5.29: Horizontal wind vectors calculated with the VAD algorithm from lidar data at the KITcube site near Masada (Fig. 3.6) for 8 November 2014. For the lowest 500 m Windcube data were used and for the layers above data from the Windtracer. Length and colours of the arrows represent wind velocity and the direction of the arrows indicate the wind direction. Data are averaged over 10 min and every sixth wind profile is shown.

5.3.3 Interpretation of the Observations

The comparison of the data measured at the two stations shows that the complex orography surrounding the Dead Sea modifies the air flow in the valley depending on its height above the valley floor and the specific orographic conditions surrounding the station. The data measured on the slope (MET1) are clearly influenced by the large-scale wind field. Forced channelling leads to a change of the wind direction from east to south-east, adjusting the flow such that it is more aligned with the valley axis. The wind direction changes to south-easterly, as the required directional change of the large-scale flow is thus smallest. However, a steady northerly flow is observed near the valley's surface, which is caused by the along-valley pressure gradient, resulting in a flow from high pressure located in the north towards lower pressure in the south. The typical clockwise rotation

of the wind with increasing height above the valley floor supports the findings of a pressure-driven channelling, and the counterclockwise angle between the easterly geostrophic wind and the valleys axis, which is approximately 80°, defines the channelled flow as a counter-current according to the definition formulated by Whiteman (2000). It was also observed that the pressure-driven channelling in the southern part of the valley (KITcube location) reached up to 350 m AMSL, whereas 35 km further to the north at 124 m AMSL (MET1) a forced channelling was observed. This is most likely caused by the changing cross sectional orography. In the south, the eastern mountains are higher than in the north, reaching up to 1150 m AMSL, thus reducing the influence of the large-scale flow on the lower part of the valley atmosphere.
Even though the synoptic influence is large, the influence of thermally driven wind systems can still be observed, as radiative forcing is strong in this region. Especially on the slopes, where the large-scale flow is only slightly channelled along the valley axis, the diurnal upslope winds interrupted the south-easterly flow during daytime, causing a sudden change to weak north-westerly flow with a reduction of the wind speed by 15 m s^{-1} (Fig. 5.27). Those upslope winds also transported moisture from the valley floor upwards. A modification of the wind field was also observed in the valley, but the wind systems induced thermally during daytime caused a positive superposition, and the lake breeze, which normally has a north-easterly direction at the western coast of the Dead Sea, considerably enhanced the pressure-driven northerly flow. During the evening and night, westerly downslope winds weakened the pressure-driven flow and wind velocities became lower. It would also be interesting to investigate the stratification and the vertical humidity distribution, but unfortunately no radiosonde measurements were available during this event.

5.4 Discussion

The combination of long-term measurements and comprehensive three-dimensional measurements during the SOPs allowed a detailed analysis of the wind systems in the Dead Sea valley. Mostly diurnal, thermally driven wind systems occur and can therefore be distinguished according to their time of occurrence on a daily scale.

The observed daytime wind systems are lake breezes and upslope winds. The slope stations show a shift in the wind direction to east directly after sunrise (Fig. 5.20), indicating the upslope winds, and the stations in the valley show a shift of the wind direction to east, when the sensible heat flux difference between the land and the water becomes positive (Fig. 5.2). The lake breeze not only dominates the wind field, it also attenuates the air temperature rise, as the cooler air from the lake is advected towards the land and slows down the diurnal temperature increase in the morning (Fig. 5.2). Similar findings were also reported by (Alpert et al., 1997) for a station south of the evaporation ponds in Sedom. The variation of the frequency of occurrence of the lake breeze on an annual scale can be attributed to the large-scale conditions. In summer, the lake breeze occurs on over 70 % of the days (Fig. 5.1) and is very persistent (Fig. 5.2). A persistent Persian Trough dominates the whole Arabian Peninsula in summer, which leads to weak pressure gradients over Israel (Goldreich, 2003). A weak large-scale forcing results in very constant and similar atmospheric conditions in the Dead Sea valley, resulting in a very persistent lake breeze. On the other hand, in winter the lake breeze can only be observed on 45 % of the days, and at half of those days a shift of the north-easterly lake breeze to south-east was observed (Fig. 5.1, P2), which is most likely caused by a superposition of the local lake breeze with the large-scale flow in winter.

DWFs were frequently observed in the evenings in the Dead Sea valley. The analysis showed that in summer, on over 60 % of the evenings, such DWFs evolved, exceeding the strength of typical, purely thermally driven downslope winds. This is quite different to other parts of the world, were downslope windstorms are also observed, but not that frequent (Zhong et al., 2008; Plavcan et al., 2014). The high probability of DWFs in the Dead Sea valley can be explained as follows: The primary trigger mechanism of DWFs in the Dead Sea valley is the evolution of a temperature gradient between the air in the valley (warm) and the air around ridge height (cooler). As a result of the very constant synoptic conditions over Israel, especially in summer, the required temperature difference can evolve frequently. It evolves either due to the prolonged warming of the air in the valley and radiative cooling at the ridge and at the slopes, or as a result of the advection of cooler maritime air towards the ridge by the Mediterranean Sea Breeze Front, which was reported by several authors, e.g. Bitan (1974), Alpert et al. (1997), Naor et al. (2015). This is quite different to downslope windstorms observed in mid-latitudes. For example in the Alps, large-scale processes, such as cold air advection connected to frontal systems, generally result in the required temperature differences between the valley and the upstream air mass, which triggers a downslope windstorm (Mayr and Armi, 2008, 2010), or large-scale processes cause mountain waves, which result in strong downslope windstorms (Jackson et al., 2013).

However, there is not only one type of DWF in the Dead Sea valley. Three types were identified according to their duration. They are rather diverse concerning their characteristics. According to the observations, short and weak events have a mean wind speed of 3 to 6 $\mathrm{m\,s^{-1}}$ and a duration of only 2 h (Tab. 5.1). Long and weak events last over 11 h, are very shallow, and have similar wind speeds like the short events (Tab. 5.1). The DWFs which lasted about 4 to 5 h had considerable higher maximum wind velocities of over 10 $\mathrm{m\,s^{-1}}$, and penetrated over 13 km into the valley (Fig. 5.9). For this type of events, a secondary forcing was found which intensified the strength of the DWF. As described with the conceptual model in Sec. 5.2.3, the transition from a subcritical to a supercritical flow, caused by an undisturbed boundary layer evolution at the ridge and an inhibited evolution upstream by the local advection of maritime air over the coastal plains, was responsible for the intensification of the DWF and the resultant hydraulic jump and rotor formation. As this transition is not necessary for the evolution of DWF and does not occur that frequently, it can explain the large differences of the DWFs regarding their strength and depth.

The DWFs also interact with the valley wind systems. If the DWF is only thermally driven and thus weak and no substantial secondary forcing exists, the thermal forcing along the north-south valley axis during nighttime can overpower the density current, leading to a northerly along-valley flow and stop the DWF penetration into the valley. This happens for the short events (Fig. 5.5). On the other hand, if the DWF is strong and can penetrate far into the valley, a total exchange of the valley air mass leads to a suppression of the radiatively induced cooling process of the valley air mass and thus no along-valley wind develops (Fig. 5.4 c, 5.11).

During nighttime northerly along-valley flows dominate. The along-valley temperature gradient, caused by different radiative cooling of the air masses along the valley axis, is stronger than the land/water contrast and thus along-valley winds dominate instead of a land breeze. The along-valley temperature gradient can be influenced by several processes. Firstly, the Topographic Amplification Factor (TAF) plays a role (see also Sec. 2.1.3) as the cross sectional area of the valley varies along the valley axis, and secondly, the southern evaporation ponds are shallower and thus warmer than the Dead Sea. This results in a stronger heat input from the water surface of the evaporation ponds to the atmosphere than over the Dead Sea. The cooling of the air in the southern part of the valley is thus delayed, resulting in the necessary along-valley temperature gradient causing the northerly along-valley flow. The height difference between the southern evaporation ponds (-395 m AMSL) and the northern Dead Sea (-430 m AMSL) is not large enough to counter act the north-south temperature gradient and thus does not influence the northerly along-valley flow. However, the heat input from the ponds to the atmosphere only delays the cooling and thus the along-valley temperature gradient weakens in the course of the night, leading to cessation of the along-valley flows. A more detailed analysis of this theory could not be performed, as no sensible heat flux measurements from the water surface of the ponds were available in 2014.

Independent of the diurnal cycle of the wind systems, large-scale forcing can overpower the diurnal, thermally driven wind systems and thus impact the valley ABL for several days. This occurs mostly in winter when the synoptic situation is more active and frontal systems pass over Israel, or when the

Red Sea Trough impacts the region. Red Sea Trough situations either cause easterly to south-easterly flow over the Dead Sea area or westerly to north-westerly flow, depending on the location of its axis. As a result of the interaction with the complex orography, strong wind channelling can occur, inhibiting the typical CBL evolution and leading to distinct atmospheric conditions in the valley.

6 Energy Balance in the Dead Sea Valley

In Sec. 5.1 it was already shown that the difference of the sensible heat flux between water and land is responsible for the lake breeze and it was shown that wind systems can considerably enhance evaporation (Sec.5.3). Now, a detailed analysis of the energy balance in the Dead Sea valley will be given to fully understand the feedback mechanisms with the wind systems and to answer the still remaining questions about the evaporation from the water surface and its predictability using high resolution eddy covariance measurements. First, results regarding the data quality and the flux footprint are presented because this is very important for the correct interpretation of eddy covariance data (Sec.6.1). Then, a regression model for the turbulent fluxes is presented in Sec. 6.2, followed by a discussion of the energy balance closure (Sec. 6.3). The diurnal and intra-annual variability of the turbulent fluxes is presented in Sec. 6.4 and the evaluation of typically used formulas to estimate evaporation on their applicability for the Dead Sea is shown in Sec. 6.5. Finally, a discussion about the impact of the results on the valley scale is given in Sec. 6.6.

6.1 Flux Data Coverage, Quality Criteria and Footprint Analysis

Data from March 2014 until March 2015 were used for the analysis. At EBS1 data for the sensible and latent heat flux were only available until 11 October 2014 due to IRGASON malfunction. At EBS2 continuous measurements were available, and at EBS3 the IRGASON was working until 19 January 2015.

Table 6.1: Rejected data due to different quality screening criteria and final data availability of sensible (H) and latent heat flux (LE) data for EBS1, EBS2, and EBS3 in the measurement period from March 2014 until March 2015.

	Malfunction (%)		Signal Strength (%)		Foken Scheme (%)		Footprint (%)		Available Data (%)	
	H	LE	H	LE	H	LE	H	LE	H	LE
EBS1	20.3*	20.2*	-	7.8	8.4	20.1	–	–	71.3	51.9
EBS2	2.1	2.4	-	10.0	11.6	9.2	19.2	19.3	67.1	59.2
EBS3	11.3+	15.2+	-	3.3	5.7	11.6	–	–	83.0	69.9

*increased because of malfunction of the IRGASON beginning at 11 October 2014.
+increased because of malfunction of the IRGASON beginning at 19 January 2014.

The data loss due to long-term and short-term malfunctions was about 20.3 % for sensible and latent heat flux at EBS1 (Tab. 6.1). At EBS2 it was only around 2.4 % and at EBS3 it was 11.3 % for sensible and 15.2 % for latent heat flux (Tab. 6.1). In addition, several quality criteria were applied to the remaining 30 min flux data. Latent heat flux data were rejected during precipitation events, when the infrared signal strength of the IRGASON was below 50 % or when the signal variability within the 30 min interval was higher than 0.6 %, assuming that this indicated disturbance of the H_2O measurements. At EBS1 7.8 % of the data were rejected due to this criteria, at EBS2 even 10 %, but at EBS3 only 3.3 % were rejected (Tab. 6.1). Half hourly flux data were also rejected when the quality flag after Foken (1999) was above class 6 (Sec. 4.2.3) and data were not already missing due to the malfunction of the instrument. Especially the quality of the latent heat flux data from EBS1 was not that good and 20 % of the data had to be removed. This leads to a final data availability of 71 % for sensible heat flux and 52 % for latent heat flux at EBS1 and 83 % and 70 % at EBS3 (Tab. 6.1). At EBS2 another criterion of the eddy covariance theory was not fulfilled. The surroundings of the station are not homogeneous. They include two main land cover types, water and bare soil (Fig. 3.4 b). Therefore, the fetch of the flux data was used as another criterion to filter the data. All flux data for wind directions between 230° and 330° were rejected, as for these wind directions the fetch was over land (Fig. 6.1). For the other wind directions the footprint analysis showed that the average source area contributing to 80 % of the fluxes was over water and ranged from about 300 m for southerly wind directions to 600 m for northerly wind directions. The rejected amount of flux data through the footprint analysis was 19 % and was mainly restricted to a short time window of about 4 h in the evening, as it was shown in Sec.5, for spring until autumn. Only in winter westerly wind conditions prevail throughout the night. The remaining data availability at EBS2 was then 67 % for sensible heat flux and 59 % for latent heat flux (Tab. 6.1). Although the data coverage for EBS2 seems quite low it is comparable to other long-term eddy covariance studies made over lakes. For comparison, Mammarella et al. (2015) reported a data coverage of 63 % for sensible heat flux and only 53 % for latent heat flux and Jonsson et al. (2008) even excluded 54 % of their flux data from a lake site. Gaps in the data set were only closed for the calculation of the daily evaporation amounts. To fill the gaps the median of the corresponding time step of the respective month was used. Uncertainties as to this approach were estimated by the median absolute deviation (MAD).

6.2 Regression Model for Turbulent Fluxes at EBS2

It was shown that the source area of the turbulent fluxes at EBS2 for wind directions between 230° and 330° is located over land due to the installation of the station at the shoreline (see Sec. 3.2.1). The amount of flux data which was rejected due to the land fetch is about 19 % (Sec. 6.1). To calculate the turbulent fluxes for these wind conditions extensive calculations were performed in order to find a suitable regression model to calculate the flux data for this wind direction sector.

First, the linear correlation of sensible and latent heat flux with several meteorological variables was investigated. The results of the variables, which had at least some correlation with the fluxes, are shown in Tab. 6.2.

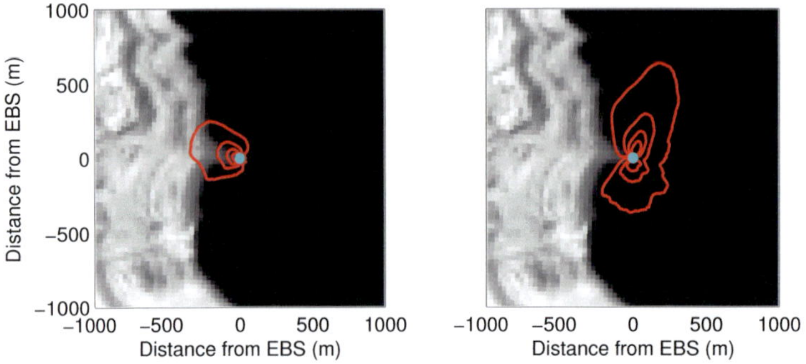

Figure 6.1: Landsat 8 image of the headland and the location of EBS2 (blue dot). Contour lines represent 20 %, 40 %, 60 %, and 80 % of the flux footprint area calculated with the footprint model of Kljun et al. (2015) for (a) offshore wind conditions with wind direction between 230° and 330° and (b) for the other wind directions. Satellite data provided by the U.S. Geological Survey.

The sensible heat flux is only weakly correlated to wind speed, with a correlation coefficient of 0.34 for the total data set and similar values for the meteorological seasons. There is no correlation to water temperature in 1 m depth ($T_{w,1\mathrm{m}}$) or to surface water temperature (T_{MO}) but a strong correlation to the temperature difference of the surface and the air temperature at 2 m. The correlation to this temperature difference explains 66 % of the variance of the sensible heat flux for the total data set and similar percentages are reached for the meteorological seasons. For the latent heat flux there is a strong correlation with wind speed, resulting in correlation coefficients of 0.53 to 0.81 for the individual meteorological seasons and 0.60 for the total data set. With water temperature in 1 m depth a correlation can only be seen in spring and autumn, where the correlation coefficient is up to 0.35. With the calculated surface water temperature (T_{MO}) a stronger correlation of 0.36 can only be found in autumn. Furthermore, the latent heat flux depends on vapour pressure deficit, Δe_{MO} and Δe_{T_w}, calculated with T_{MO} and with $T_{w,1\mathrm{m}}$, respectively. In autumn and winter a strong correlation

of 0.46 and 0.56 for Δe_{MO} and 0.40 and 0.45 for Δe_{T_w} can be found, and for the total data set it is 0.38 and 0.35, respectively.
Although there are already some good correlations with individual variables, none of them can fully explain the variance of the sensible and latent heat flux. Therefore, a stepwise linear regression model (Sec. 4.4) was applied to find the best fitting solutions for the turbulent fluxes of the water surface. For the latent and the sensible heat flux two variable sets for each flux were tested as predictor variables:

$$X_{LE,1} = (v_a, \Delta e_{T_w}, T_{w,1\text{m}}, T_{MO}) \tag{6.1}$$

$$X_{LE,2} = (v_a, \Delta e_{T_{MO}}, T_{w,1\text{m}}, T_{MO}) \tag{6.2}$$

$$X_{H,1} = (v_a, \Delta T_{w-2\text{m}}, T_{w,1\text{m}}, T_{MO}) \tag{6.3}$$

$$X_{H,2} = (v_a, \Delta T_{MO-2\text{m}}, T_{w,1\text{m}}, T_{MO}) \tag{6.4}$$

The choice of the variables was based on the results of the linear regressions in Tab. 6.2. The stepwise linear regression was performed for each variable set separately, for each meteorological season, and for the whole data set. For the variable set $X_{LE,1}$ the stepwise linear regression model gives different dependencies of the latent heat flux on the predictor variables for the different seasons. For the summer season, only wind speed is considered as a relevant variable for the regression of the latent heat flux, which results in a correlation coefficient of 0.72 (Tab. 6.3). In spring and winter, the model considers wind speed and Δe_{T_w} as relevant and the correlation is better with $R = 0.81$ in spring and $R = 0.86$ in winter. In autumn and for the total data set the surface water temperature is considered as relevant additionally, resulting in correlation coefficients of 0.84 and 0.83, respectively. With the variable set $X_{LE,2}$ the latent heat flux depends on wind speed and vapour pressure deficit in all meteorological seasons (Tab. 6.3), and the correlation coefficient varies between 0.77 and 0.85. It is important to note that $T_{w,1\text{m}}$ is not added to the model in any of the cases. This is due to the fact that evaporation takes place directly at the water surface in

Table 6.2: Correlation coefficients of sensible heat flux (H) and latent heat flux (LE) with wind speed (v_a), water temperature in 1 m depth ($T_{w,1m}$), surface water temperature (T_{MO}), vapour pressure deficit calculated with 1 m water temperature (Δe_{T_w}), and with surface water temperature ($\Delta e_{T_{MO}}$), temperature difference between surface water temperature and 2 m air temperature (ΔT_{MO-2m}), and difference between water temperature at 1 m depth and 2 m air temperature (ΔT_{w-2m}) for the meteorological seasons and the total data set.

	v_a		$T_{w,1m}$		T_{MO}		Δe_{T_w}		$\Delta e_{T_{MO}}$		ΔT_{MO-2m}		ΔT_{w-2m}	
	H	LE	H	LE	H	LE	H	LE	H	LE	H	LE	H	LE
Spring	0.39	0.68	-0.08	0.23	-0.19	0.07	–	0.24	–	0.06	0.83	0.17	0.61	0.23
Summer	0.35	0.72	-0.13	0.02	-0.02	0.00	–	-0.09	–	-0.12	0.76	-0.07	0.44	-0.06
Autumn	0.34	0.53	0.05	0.35	0.11	0.36	–	0.40	–	0.46	0.78	0.01	0.19	-0.23
Winter	0.43	0.81	-0.02	0.08	0.18	0.19	–	0.45	–	0.56	0.87	0.25	0.39	0.06
Total	0.34	0.60	-0.09	0.43	-0.09	0.42	–	0.35	–	0.38	0.81	0.00	0.38	-0.18

Table 6.3: Results of the stepwise linear regression model for the latent heat flux of the water surface with variable set $X_{LE,1}$ and $X_{LE,2}$. The corresponding correlation coefficient (R) of the model is shown if a variable is added to the model. If a variable is not added to the model, it is indicated by a minus sign. The prediction errors yielded by the MCCV with randomly chosen validation data points (er_r) and randomly chosen validation sectors (er_s) are given. Results are shown for the meteorological seasons and for the whole data set.

	$X_{LE,1}$					$X_{LE,2}$				
	v_a	Δe_{T_w}	T_{MO}	er_r (%)	er_s (%)	v_a	$\Delta e_{T_{MO}}$	T_{MO}	er_r (%)	er_s (%)
Spring	0.69	0.81	–	1.25	8.31	0.68	0.77	–	0.50	8.00
Summer	0.72	–	–	0.82	-1.09	0.73	0.77	–	0.27	1.64
Autumn	0.55	0.79	0.84	0.04	6.17	0.53	0.82	–	0.12	2.30
Winter	0.80	0.86	–	3.19	15.30	0.81	0.85	–	2.96	1.36
Total	0.59	0.78	0.83	0.71	5.94	0.59	0.80	–	1.02	5.01

a layer of approximately 10 µm (Emery et al., 2001). The water temperature in 1 m depth does not represent the temperature of that thin surface layer and its short-term changes (Nehorai et al., 2013), and is hence not suitable for the regression with the latent heat flux. Besides the correlation coefficients, the robustness and the accuracy of the models were tested with a Monte-Carlo cross validation (MCCV). The work flow of the MCCV and the splitting approaches used for the MCCV - (i) randomly chosen data points and (ii) randomly chosen wind direction sectors - are explained in Sec. 4.4. In this work the MCCV was performed 100 times for each splitting approach, and the prediction error (er) is then given by the mean error of the individual realisations for (i) as er_r and for (ii) as er_s. The prediction error for the latent heat flux depends on the approach used to split the data set into training and validation data sets. For the MCCV with randomly chosen data points as validation data, both models ($X_{LE,1}$ and $X_{LE,2}$) performed well for the different meteorological seasons, with a prediction error (er_r) between 0.04 % and 3.19 % for $X_{LE,1}$ and deviations between 0.12 % and 2.96 % for $X_{LE,2}$. However, for the MCCV with randomly chosen validation sectors, the model with variable set $X_{LE,1}$ overestimated the latent heat flux by

15.30 % in winter, 8.31 % in spring, and 6.17 % in autumn. The second model was more reliable. It also overestimated evaporation in spring by 8.0 %, but in the other seasons the error was only 2.3 % or even less. Although correlation coefficients of the latent heat flux model are similar for $X_{LE,1}$ and $X_{LE,2}$, the MCCV showed that the model with variable set $X_{LE,2}$ was more reliable as the error between estimated and measured flux was smaller and should therefore be preferred for estimating the latent heat flux of the water surface.

Table 6.4: Results of the stepwise linear regression model for the sensible heat flux of the water surface with variable set $X_{H,1}$, $X_{H,2}$ (see Tab. 6.3 for details).

	$X_{H,1}$					$X_{H,2}$				
	$\Delta T_{w-2\text{m}}$	v_a	$v_a{\cdot}\Delta T_{w-2\text{m}}$	er_r (%)	er_s (%)	$\Delta T_{MO-2\text{m}}$	v_a	$v_a{\cdot}\Delta T_{MO-2\text{m}}$	er_r (%)	er_s (%)
Spring	0.63	0.70	0.79	3.21	17.41	0.83	0.90	0.95	6.41	8.59
Summer	0.47	0.58	–	8.08	18.78	0.78	0.87	0.93	7.38	6.22
Autumn	0.34	0.43	0.55	3.49	41.56	0.80	0.87	0.93	7.06	0.87
Winter	0.43	0.61	0.68	-5.92	46.47	0.87	0.90	–	7.20	-5.86
Total	0.41	0.56	0.63	2.91	35.99	0.82	0.88	0.93	9.22	4.61

The model of the sensible heat flux showed strong differences between variable set $X_{H,1}$ and $X_{H,2}$. For $X_{H,1}$ the correlation coefficients were quite low and varied between 0.55 and 0.79. As predictor variables, $\Delta T_{w-2\text{m}}$, wind speed, and the product of wind speed and $\Delta T_{w-2\text{m}}$ were considered as relevant in all cases except for the summer (Tab. 6.4). For $X_{H,2}$ also $\Delta T_{MO-2\text{m}}$, wind speed, and the product of wind speed and $\Delta T_{MO-2\text{m}}$ were considered as relevant parameters in all cases except for winter, but correlation coefficients were higher than for $X_{H,1}$, varying between 0.93 and 0.95 (Tab. 6.4). The prediction error for the sensible heat flux also depended on the approach used to create the validation data sets. For the MCCV with randomly chosen data points as validation data, model $X_{H,1}$ had an error ranging from -5.92 % in winter to 8.08 % in summer, and model $X_{H,2}$ overestimated the sensible heat flux in all season by 6.41 to 9.22 %. On the other hand, for the MCCV with randomly chosen validation sectors, the model with the variable set $X_{H,1}$ produced errors between 17.4 and 46.47 %. This means that the model depends on the data used to train the model. In this case the training data were not representative for the validation data set. For the second variable set, $X_{H,2}$, the model worked better and the spread between -5.86 and 8.59 %, which is still an acceptable error range for the model. The smaller model error for $X_{H,2}$ together with much better regression coefficients lead to the conclusion that this model has to be used to estimate sensible heat flux from the water surface.

For the further analysis of the flux data, the regression model with $X_{LE,2}$ for latent heat flux and $X_{H,2}$ for sensible heat flux is applied to the data from EBS2. With the regression model, 82 % of the sensible heat flux data and 90 % of the latent heat flux data, which were rejected due to the fetch conditions, could be calculated, leading to a data availability of 82.8 % for the sensible and 76.8 % for the latent heat flux after the regression model was applied. This is quite important for the analysis of the diurnal and intra-annual variation of the turbulent fluxes. Especially for the latent heat flux, as can be seen in Fig. 6.2. The measured (uncorrected) fluxes and the corrected fluxes are similar during the day, as during this time the local lake breeze prevails and nearly no westerly wind directions occur (Sec. 5.1). Hence nearly no fluxes have to be calculated with the regression model. On the other hand, in the evening between 18 and 22 LT, very high differences can be seen in spring, summer, and autumn (Fig. 6.2), caused by the already discussed westerly DWFs (Sec. 5.1). The uncorrected flux represents the latent heat flux from the land surface during this time interval, and the corrected flux represents the latent heat flux from the water surface. The differences are 73 W m^{-2} in spring, 150 W m^{-2} in summer, and 100 W m^{-2} in autumn (Fig. 6.2). The corrected flux values for the water surface during this time frame are even higher than the values during daytime. This can be explained by the high wind speeds connected to the DWFs and by the exchange of the valley air mass with drier air from outside of the valley causing an increase of the vapour pressure deficit (Sec. 5.2). As the latent heat flux depends on wind speed and the vapour pressure deficit (Tab. 6.3), which both increase with the onset of the DWF, the latent heat flux increases correspondingly. Without the regression model it would not be possible to describe these extreme values in the evening and hence the diurnal cycle of the latent heat flux would not be correct. For the sensible heat flux, the differences between the uncorrected and corrected values are quite small with values of only 5 W m^{-2}. This is due to the fact that the sensible heat flux depends only slightly on wind speed and is mainly controlled by the temperature difference between the surface water temperature and the air temperature (Tab. 6.4). This variable is nearly not influenced by the DWFs and hence does not change the sensible heat flux considerably.

Table 6.5: Energy partitioning for the meteorological seasons at all EBSs and energy balance closure (EBC) for EBS1 and EBS3. For EBS2 ΔQ is calculated as the residual value of Rn and the turbulent fluxes ($H+LE$). At EBS1 no results for the winter months could be calculated due to the instrument malfunction (see Sec.6.1).

	H/Rn			LE/Rn			G/Rn	$\Delta Q/Rn$	G/Rn	EBC (%)	
	EBS1	EBS2	EBS3	EBS1	EBS2	EBS3	EBS1	EBS2	EBS3	EBS1	EBS3
Spring	0.58	0.06	0.29	0.17	0.43	0.75	0.05	0.51	0.02	80	106
Summer	0.70	0.05	0.26	0.09	0.53	0.76	0.06	0.42	0.06	85	108
Autumn	0.80	0.10	0.14	0.16	0.74	0.91	-0.03	0.16	-0.04	93	101
Winter	–	0.39	0.03	–	0.92	0.90	–	-0.31	-0.14	–	80

6.3 Energy Balance Closure

The energy balance closure (EBC) is a consequence of the first law of thermodynamics and states that the sum of the turbulent fluxes ($H+LE$) has to be equal to the source and sink terms ($Rn-G$) (Wilson et al., 2002). If there is an imbalance of the EBC, then either the turbulent fluxes are under- or overestimated or there is an error in the source and sink terms. The analysis of the EBC gives an assessment of the quality of the eddy covariance results (Anderson et al., 1984; Mahrt, 1998) and gives implications on how the results of the

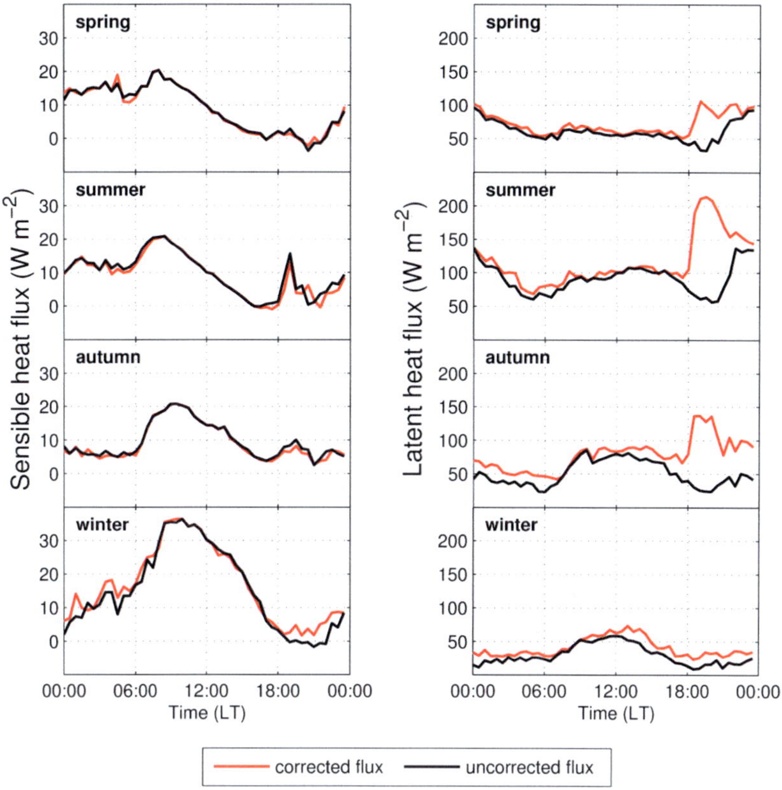

Figure 6.2: Median diurnal cycles of the actually measured turbulent fluxes (black lines) at EBS2 and the fluxes corrected with the multiple regression model for wind directions between 230 and 330°. The corrected values represent the turbulent fluxes from the water surface independent of the wind direction.

turbulent fluxes should be interpreted (Wilson et al., 2002). Several reasons for a lack of the EBC have been discussed in literature over the past 30 years and were recently summed up in an overview paper from Foken (2008). The reasons which should still be considered when analysing the EBC are, after Foken et al. (2012b), the heat storage in the soil between the sensor and the surface and fluxes connected to larger eddies, which can not be measured with the eddy covariance technique. Recent studies showed that spatially averaged fluxes (Inagaki et al., 2006; Kanda et al., 2004) or temporally averaged fluxes over longer time scales (if the stationarity assumption is still fulfilled) (Sakai et al., 2001; Finnigan et al., 2003) could close the energy balance.

For the station over bare soil (EBS1) and the station over vegetation (EBS3) the EBC for the meteorological seasons and the relative contribution of the sensible, latent and ground heat flux on the net radiation partitioning was calculated (Tab. 6.5). For EBS2, the station at the shoreline, no direct measurements for the calculation of the heat storage term of the water body were available. Therefore, only the relative contribution of the turbulent fluxes on the net radiation partitioning was calculated and the residual was assumed to be the heat storage term

$$\Delta Q = Rn - H - LE. \tag{6.5}$$

The net radiation for the water surface was calculated using Eq. 4.21.

Over bare soil (EBS1) between 58 % and 80 % of the available net radiation was transformed into sensible heat from spring until autumn. Only a minor part (9 % - 17 %) of the net radiation was transformed into latent heat and even less was measured for the ground heat flux. This leads to an EBC over bare soil of 80 % to 93 % from spring until autumn and is in the range of other energy balance studies, e.g. Leuning et al. (1982); Kanemasu et al. (1992); Wilson et al. (2002), who found a general lack of the energy balance closure of about 20 % in their studies. The most likely reason for the imbalance at EBS1 is an underestimation of the ground heat flux. Heusinkveld et al. (2004) investigated the role of ground heat flux on EBC in the Negev (Israel) and found that the dry desert soil in the Negev desert had a damping depth of only 64 mm (compared to damping depths for moist soils in mid latitudes of around 200-300 mm). This means that the diurnal surface temperature range was already reduced to 1/e at 64 mm depth, resulting in large differences between the ground heat flux directly at the surface and in standard measurement depths of about 40 to 50 mm. Heusinkveld et al. (2004) found differences of up to 100 W m^{-2} between measurements at 46 mm depth and those made at 1 mm depth. With his measurements at 1 mm depth he was able to achieve an EBC of 96 %. At EBS1 the soil is also very dry and, therefore, a similar damping depth as in the study of Heusinkveld et al. (2004) is realistic. This accounts most likely for most of the energy balance residual at this site. Another possible measurement problem could be caused by cavities between the sensor and the soil, which serve as an additional isolation, and therefore influence the measurements. Even though the sensors were installed with uppermost caution, such cavities could probably not be avoided, due to the dry soil. Of course, large scale eddies, as proposed in Foken (2008), can not be neglected as a reason for the imbalance of the EBC, but it can not be investigated with the available data set.

At the water surface (EBS2), about 43 % of the energy was converted into latent heat flux in spring and about 57 % in summer. The sensible heat flux accounted for only 6 % and 5 % of the net radiation in these seasons, respectively, leading to a high heat storage amount of about 51 % and 42 % in spring and summer (Tab. 6.5). In autumn the heat storage term was only 16 % as the turbulent fluxes accounted for 84 % of the available net radiation. In winter, the accumulated heat in the water column was released, compensating the turbulent fluxes, which exceeded the available net radiation by 31 %.

At EBS3 most of the net radiation was transformed into latent heat flux, which was 75 % in spring and summer and even 90 % in autumn and winter. The sensible heat flux accounted for 29 % in spring, 26 % in summer, 14 % in autumn and only 3 % in winter. The energy balance closure at EBS3 was quite good from spring until autumn, where the sum of sensible, latent and ground heat flux slightly overcompensated the available net radiation by 6 %, 8 % and 1 %, respectively. In winter the turbulent fluxes also accounted for 93 % of the net radiation, but a relatively large negative ground heat flux, meaning that energy was transported towards the surface, resulted in an imbalance of the energy balance by 20 %. Overall, the results of the EBC analysis show that the eddy covariance measurements at the sites are representative for the turbulent fluxes. The imbalance at EBS1 is in the range of other EBC studies and results most likely from the underestimation of the ground heat flux and maybe from some larger eddies which can not be measured by a single EBS, and not from a systematic underestimation of the turbulent fluxes. Also at EBS3 the measured turbulent fluxes are representative for this measurement site, and the good energy balance closure results most likely from the governing influence of the canopy reducing the impact of the ground heat flux compared to EBS1.

6.4 Diurnal Variation

After the analysis of the EBC at the different sites the diurnal variation of the energy fluxes of the different surfaces will be discussed. Fig. 6.3 shows the median diurnal cycles of the fluxes. The net radiation of the water surface is again calculated with Eq. 4.21.

Net Radiation

Net radiation peaked around local noon at all sites and in all seasons (Fig. 6.3) and reached values of up to 513 W m^{-2} at EBS1, 845 W m^{-2} at the water surface and 597 W m^{-2} at EBS3 in summer. Not only the maximum values were largest at EBS2, but also the total amounts of net radiation (Tab. 6.6). The total amounts correspond to the sum over the median cycles and are twice as high at the water surface as over bare soil (EBS1) for spring until autumn (Tab. 6.6). Also in winter net radiation at the water surface is higher than over bare soil, but the ratio between the net radiation at EBS1 and at the water surface increases to 0.72. The ratio of net radiation at EBS1 and EBS3 is nearly constant throughout the year, with values around 0.75 for spring, summer and winter, and 0.69 in autumn. Comparing the net radiation between EBS3 and the water surface yields ratios of 0.69 for spring and summer, 0.77 for autumn, and is nearly 1 in winter. This means that in all seasons the lowest energy amounts are available at EBS1 and that the most energy is available at the water surface, but in winter nearly the same amount is available at EBS3 as well. The differences of the net radiation amounts between the sites were caused by two factors. First, the albedo of the

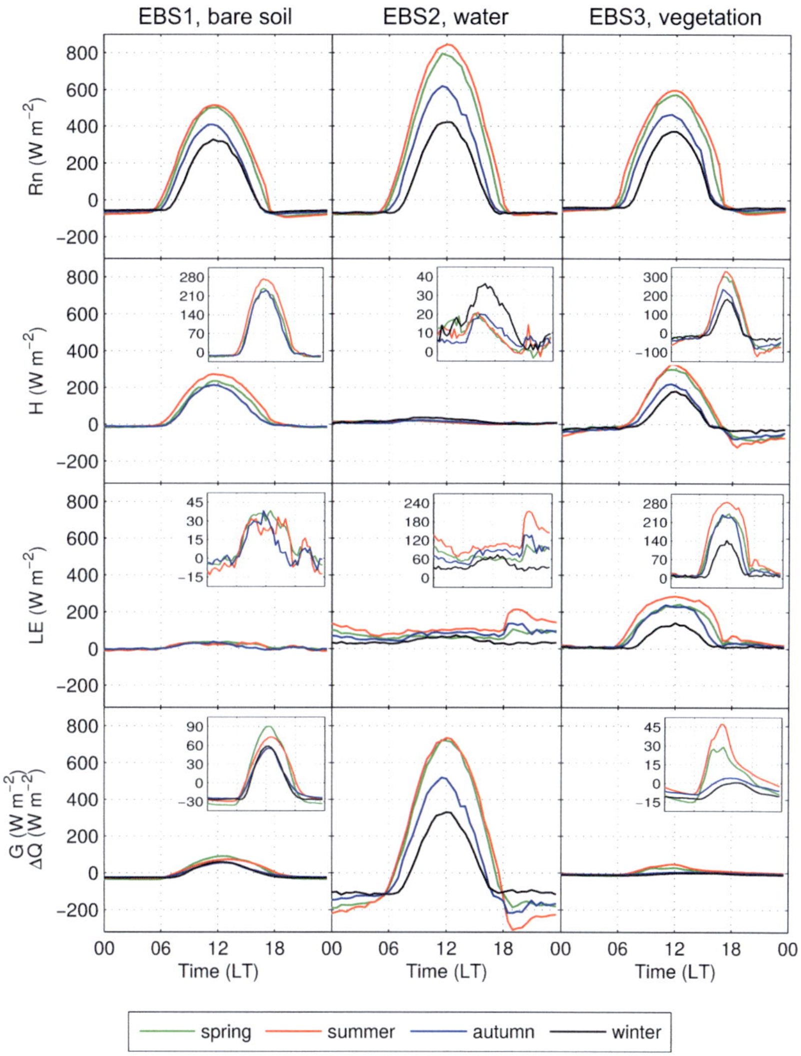

Figure 6.3: Median diurnal cycle of energy balance components for the meteorological seasons 2014/15. From top to bottom: net radiation, sensible heat flux, latent heat flux, and ground heat flux for EBS1 and EBS3, and heat storage term for EBS2 are shown. All axes have the same range for better comparison and the inserts show the data with an adjusted y-axis according to the specific value range of the variable.

Table 6.6: Median daily amount of net radiation at the three sites and the ratio of the net radiation between the sites for the meteorological seasons.

	$Rn\,(\mathrm{W\,m^{-2}})$			Ratio of Rn		
	EBS1	EBS2	EBS3	EBS1/EBS2	EBS3/EBS2	EBS1/EBS3
Spring	4430	8475	5828	0.52	0.69	0.76
Summer	5469	10662	7409	0.51	0.69	0.74
Autumn	2841	5394	4145	0.53	0.77	0.69
Winter	1836	2560	2460	0.72	0.96	0.75

surface was different and thus the reflected shortwave radiation. At EBS1 the albedo was highest, with $\alpha \approx 0.27$, leading to the highest amount of outgoing shortwave radiation, followed by EBS3 with an albedo of $\alpha \approx 0.22$, and the smallest albedo was at the water surface, with $\alpha \approx 0.07$, leading to very small amounts of reflected shortwave radiation (Fig. 6.4). The second reason for the low net radiation at EBS1 was a very high surface temperature during the day, leading to high amounts of upward longwave radiation, as this is proportional to the fourth power of the surface temperature (Eq. 2.15). Surface temperature at EBS1 reached maximum values of around 53 °C in summer, whereas at EBS3 only 45 °C were reached, and surface water temperature at EBS2 peaked at around 37 °C, leading to considerably higher longwave upward radiation fluxes at EBS1 and EBS3 compared to EBS2 (Fig. 6.4). The slightly lower surface temperature and the lower albedo at EBS3 resulted most likely from the vegetation at this site. Firstly, the surface directly below the radiometer was not completely free from vegetation, leading to a lower albedo, and the transpiration of the plants also reduced their surface temperatures. Secondly, the reed around EBS3 casted shadow over the surface from time to time and thus reduced the surface heating. In winter the surface temperature at EBS1 and EBS3 were comparable to the surface water temperature, resulting in similar amounts of upward longwave radiation (Fig. 6.4) and therefore smaller differences between the net radiation at the water surface and at the other stations.

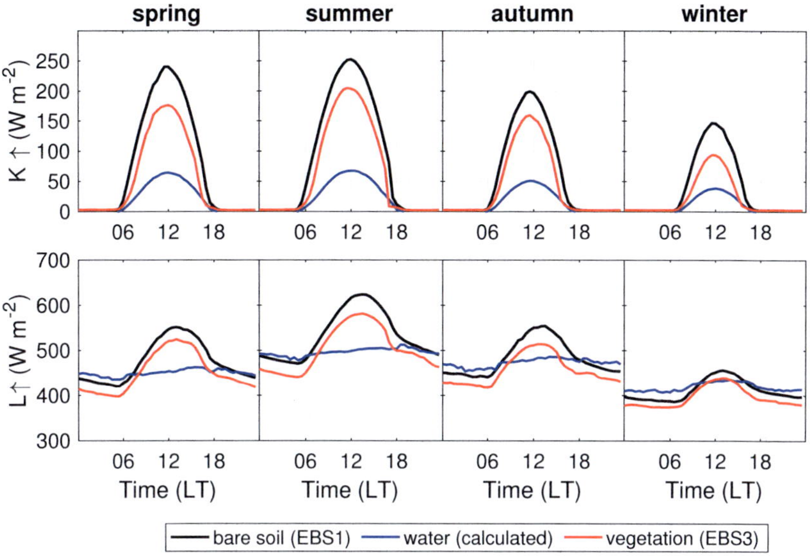

Figure 6.4: Median diurnal cycle of the upward shortwave and longwave radiation fluxes at EBS1, at the water surface, and EBS3. The shortwave radiation flux at the water surface was calculated with an albedo of 0.07, and the longwave radiation flux was calculated using the surface water temperature.

Sensible Heat Flux

The diurnal cycle of the sensible heat flux is very different at the three sites. At EBS1 it peaked around noon, just like the net radiation, and reached values of 273 W m^{-2} in summer (Fig. 6.3). In spring and autumn the maximum values were slightly smaller. During the night a small negative sensible heat flux, which was on average around -16 W m^{-2}, was observed from spring until autumn. In comparison to the diurnal cycle of the net radiation a delayed increase in the morning and a delayed decrease in the evening could be observed. This was probably caused by the increased latent heat flux in the morning (Fig. 6.3).

The sensible heat flux from the water (EBS2) reached its maximum in the morning around 9:00 LT and its minimum in the late evening around 21:00 LT in all meteorological seasons. The peak-to-peak amplitude ranged from 17 W m^{-2} in summer to 39 W m^{-2} in winter (Fig. 6.3).

In the morning, at EBS3, sensible heat flux started to increase about 1 h later compared to the net radiation, but peaked around noon as well. During the evening it was strongly negative (Fig. 6.3). The peak-to-peak amplitude in winter was 222 W m^{-2} and reached 452 W m^{-2} in summer. The strong negative sensible heat flux during the night was caused by an oasis effect. Drier air is advected towards the vegetated site with enhanced wind speeds, causing an increase of the evaporation and thus the latent heat flux, which leads to a negative sensible heat flux. The process chain of such an event is illustrated with Fig. 6.5 showing the diurnal cycle of the relevant meteorological variables for the 26 August 2014. With a sudden wind direction change at 16:30 LT wind speed increases from 0.5 to 4 m s^{-1} and specific humidity drops by 3.5 g kg^{-1}. At the same time the latent heat flux increases strongly from 20 to 260 W m^{-2} even though net radiation is already negative. The energy necessary for the evaporation process is provided by a negative sensible heat flux of the same amount, leading to a decrease in air temperature.

Latent Heat Flux

The latent heat flux over bare soil was very small, with a maximum of about 35 W m^{-2} around noon in spring and summer and around 9:00 LT in autumn (Fig. 6.3). During the night latent heat flux was negative, indicating dew deposition, which is typical for arid environments. The peak-to-peak amplitude was around 45 W m^{-2} for spring until autumn (Fig. 6.3). With such low latent heat flux values the Bowen ratio reached $\beta \approx 12$ for midday in summer, which seems reasonable for the arid conditions in the Dead Sea valley. The value is comparable to other energy balance studies in arid environments. Kalthoff et al. (2006) reported Bowen ratios of $\beta \approx 11$ for an arid valley in Chile, Krishnan et al. (2012) calculated Bowen ratios of $\beta \approx 10$, and Kampf and Tyler (2006) found Bowen ratios of around 10 for alluvial sediments in the Salar de Atacama in Chile.

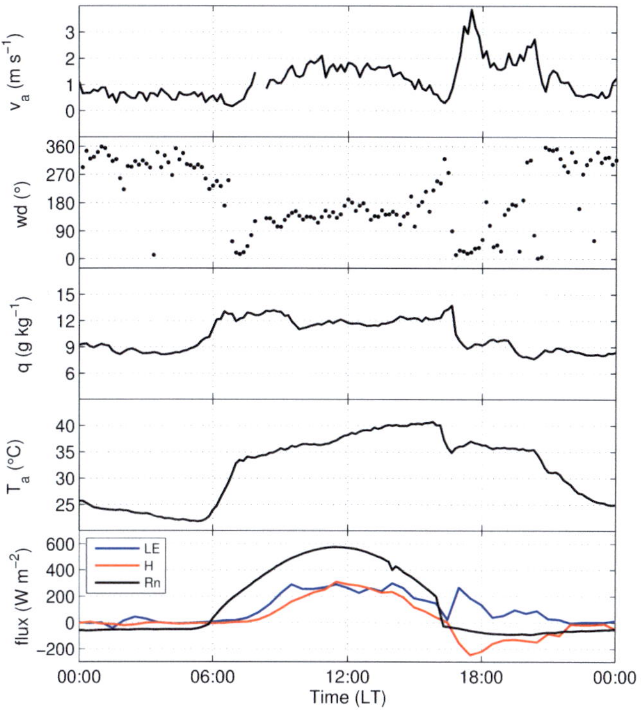

Figure 6.5: Time series of the relevant meteorological variables explaining the oasis effect for the 26 August 2014. From top to bottom: wind speed, wind direction, specific humidity, and the latent heat flux, sensible heat flux and net radiation.

For the water surface the latent heat flux was the dominant turbulent flux. Between spring and autumn the maximum was reached around 19:30 LT, when wind speed was highest (Fig. 5.2), and minimum values were found in the early morning hours around 6:00 LT together with the lowest wind speed values (Fig. 6.3). During the winter season the maximum was reached around midday, also connected to the diurnal cycle of the wind speed (Fig. 5.2). The peak-to-peak amplitude of the latent heat flux varied from only 50 W m^{-2} in winter to 145 W m^{-2} in summer.

At EBS3 the diurnal cycle of the latent heat flux was linked to the diurnal cycle of the net radiation. The latent heat flux increased strongly in the morning and reached its maximum around noon with maximum values of 285 W m^{-2} in summer (Fig. 6.3). The strong increase in the morning accounted for almost all energy of the net radiation, causing a delayed increase of the sensible heat flux in the morning (Fig. 6.3). As already mentioned before, a secondary maximum appeared in the evenings from spring until autumn through an oasis effect and minimum values were found just before sunrise. The peak-to-peak amplitude varied between 140 W m^{-2} in winter and 277 W m^{-2} in summer (Fig. 6.3). As soil moisture was comparable to EBS1, only 10 % of the latent heat flux resulted from soil evaporation. The rest of the measured latent heat flux can be attributed to the transpiration of the plants, as they get the water from deeper soil layers. This explained the diurnal cycle but also the seasonal variation. As soon as sunlight is available, the plants start their photosynthesis. They open their stomata, which leads to transpiration. Therefore, the diurnal cycle is closely related to the solar cycle. Additionally, plant activity is lowest in winter, resulting in low latent heat flux values, and highest in summer, resulting in accelerated latent heat flux values (Fig. 6.3). The Bowen ratio of $\beta \approx 1.2$ is comparable to other vegetated sites.

Ground Heat Flux

The ground heat flux at EBS1 peaked at about 13:00 LT, with maximum values of 85 W m^{-2} in spring, and has a peak-to-peak amplitude of 127 W m^{-2} (Fig. 6.3). Slightly smaller maximum values and peak-to-peak amplitudes were reached in the other seasons. As already discussed in Sec. 6.3 the ground heat flux measured at this site seemed too low for the non vegetated, arid conditions at the station. Heusinkveld et al. (2004) found for a comparable site in the Negev a very shallow damping depth of the temperature amplitude of only 64 mm (compared to 200 to 300 mm in mid-latitudes), resulting in an insufficient representation of the ground heat flux at the surface with measurements conducted at 50 mm depth.

For the energy balance of a water surface the ground heat flux is defined as the flux between the water and the soil at the bottom of the lake and is negligible for deep lakes like the Dead Sea. A more important term of the energy balance of the water surface is the heat storage term (ΔQ) (Sec. 2.3). Therefore, it is calculated as the residuum of the turbulent fluxes and the net radiation (Eq. 6.5) and is shown in Fig. 6.3. Most of the net radiation was stored during the day, peaking around noon with values of up to 720 W m^{-2} in spring and summer, and 315 W m^{-2} in winter. During the night heat was released from the water to provide the energy for the positive turbulent fluxes, especially the high latent heat flux. The heat storage term reached minimum values around 19:00 LT when the latent heat flux reached its maximum (Fig. 6.3).

At the vegetated site, very low values of ground heat flux were measured. In spring and summer, daytime ground heat flux values of around 30 to 45 W m^{-2} were measured, which corresponds to 6 % of the net radiation (Fig. 6.3). This is comparable to values reported by Stull (1988), who found ground heat flux values of about 5-15 % of the net radiation during daytime. The reason for this relatively low ground heat flux values is the plant cover, and at our site the additional shading of the ground by the reed. This reduces the direct irradiation of the ground and thus the heating of the ground and the ground heat flux. This is also supported by the median soil temperature, which ranged in summer between 27 and 38 °C at EBS3 compared to 32 and 50 °C at EBS1. In autumn the ground heat flux was considerably lower than in spring and summer and was mainly negative throughout the day. Only in the afternoon it became positive, with maximum values of 5 W m^{-2}. In winter it was even lower than in autumn, meaning that the ground heat flux was mainly directed towards the Earth's surface, serving as a source of energy. The strong reduction of the ground heat flux in autumn and winter compared to spring and summer, results mainly from an enhanced vegetation cover of the ground during this seasons and from the shading of the sensors over the whole day.

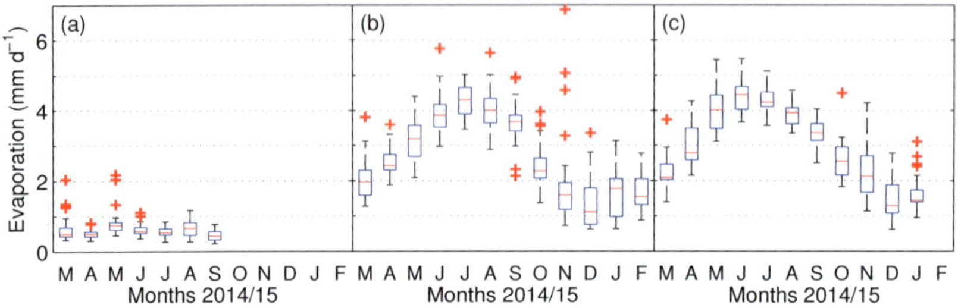

Figure 6.6: Boxplot of daily evaporation at (a) EBS1 for evaporation from bare soil, (b) EBS2 for evaporation from the water surface and (c) EBS3 for evaporation from vegetation. Red lines mark medians, the edges of the box are the 25th and 75th percentiles, the whiskers extend to the most extreme data points not considered outliers, and outliers are plotted individually by red crosses.

6.5 Daily Evaporation and Dew Amounts

6.5.1 Evaporation

The median daily evaporation at EBS1 was nearly constant throughout the measurement period. The median varied between 0.42 mm d^{-1} and 0.74 mm d^{-1} (Fig. 6.6 a). The slightly higher evaporation values in May and the outliers seen in Fig. 6.6 (a) were related to precipitation events. Generally the soil in arid regions is very dry and therefore limits the amount of evaporation as a result of the limited water supply. Only after intense precipitation events soil moisture increases and more water is available for evaporation. Evaporation rates of over 2 mm d^{-1} were therefore only reached after strong precipitation events and when soil moisture decreased after a couple of days, evaporation decreased as well. For the measured period between 1 March and 30 September 2014 an evaporation amount of 130 mm was measured.

At EBS2 the evaporation amounts were much higher than at EBS1. In spring, evaporation values steadily increased until a maximum median evaporation of 4.3 mm d^{-1} was reached in July (Fig. 6.6 b). Afterwards, evaporation values decreased until a minimum median evaporation of 1.1 mm d^{-1} was reached in December (Fig. 6.6 b). Aggregating the evaporation over the whole measurement period resulted in a total amount of 994.5 ± 81.2 mm. This is quite close to the evaporation amounts calculated by Stanhill (1994) (1000 mm) or Lensky et al. (2005) (1100-1200 mm), who estimated the evaporation based on theoretical energy balance approaches. Also visible in Fig. 6.6 (b) is the higher variation of the daily evaporation amounts between November and February. This is the so-called wet season, when synoptic patterns gain more influence on the atmospheric conditions in the valley (Bitan, 1974, 1976). The governing factors of evaporation, like wind speed and vapour pressure deficit, are very variable during this time. On the one hand, winter storms with rain and high air humidity can reach the region, which decreases the evaporation rate. On the other hand, winter storms without rain but high wind velocities, which advect very dry air to the Dead Sea (e.g. dust storms), can considerably increase the evaporation rate (Shafir and Alpert, 2011). The highest variability (not considering outliers) can be seen in January, with daily evaporation amounts between 0.64 and 3.14 mm d^{-1}. In November, evaporation values varied between 0.73 and 2.42 mm d^{-1}, but on three consecutive days evaporation exceeded these values. Evaporation amounts of 5.06 mm d^{-1}, 6.86 mm d^{-1}, and 4.57 mm d^{-1} were measured, which is the absolute maximum of the whole measurement period. These extreme evaporation values were caused by a Central Red Sea Trough (RST) and a dominant high to the east, which caused southeasterly winds above the valley. It could be observed that through the complex orography the wind was channelled along the valley axes, and a northerly wind with constantly high averaged speed of over 10 m s^{-1} advected warm and very dry air over the lake, which increased the evaporation dramatically (Sec. 5.3).

The median daily evapotranspiration at EBS3 also showed a clear annual cycle, with a maximum of 4.4 mm d^{-1} in June (Fig. 6.6 c). Afterwards, evapotranspiration decreased slowly and approached its minimum of 1.29 mm d^{-1} in December. As the measured evapotranspiration at this site is mainly caused by the transpiration of the reed, the annual cycle is connected to the growing cycle of the plants. In spring, new leafs grow and contribute to the evapotranspiration, increasing its rate rapidly. In June and July, median evapotranspiration rates of about 4.45 and 4.25 mm d^{-1} were measured, and with the beginning of the leaf senescence a reduction of the evapotranspiration was observed in August. In the consecutive months the evapotranspiration steadily decreased until the minimum was reached in December. A total annual evapotranspiration amount could not be calculated directly because data from February 2015 were missing, but as evapotranspiration is directly connected to the growing cycle of the plants and follows a steady increase in spring. The assumption of taking the average of January and March is reasonable. This resulted in an evapotranspiration amount of 47 mm for February 2015. With the measured amount of 1014 mm for the other 11 months, the total evapotranspiration was 1062 mm for the whole year.

6.5.2 Dew Deposition

Dew deposition, even if the amounts are small, is an important water source for plants in semi-arid and arid regions, as it is a rather reliable and constant source throughout the year (Jacobs et al., 1999; Malek et al., 1999; Kidron, 2000). Of the two stations on land (EBS1 and EBS3) only at EBS1 dew deposition could be observed. At EBS3 the measured latent heat flux was dominated by the transpiration of the plants. Besides the stomatal transpiration, which is closely connected to the diurnal cycle of solar radiation, plants do also transpire small amounts of water over the whole leaf surface, a process known as cuticular transpiration. This transpiration can not be controlled by the plant and is also present during the night. Therefore, the latent heat flux measured at EBS3 is also slightly positive during the night (Fig. 6.3). Daily dew deposition rates at EBS1 were calculated by aggregating negative latent heat flux values for each day. The median of the daily measured dew amount varied between 0.04 and 0.28 mm d^{-1} from March until September (Fig. 6.7). The maximum of 0.28 mm d^{-1} was reached in August. These amounts are in the range of values given by Kalthoff et al. (2006); Kidron et al. (2002); Jacobs et al. (2002); Malek et al. (1999) for other arid or semi-arid areas. The total dew amount between 1 March and 30 September 2014 was 36.2 mm

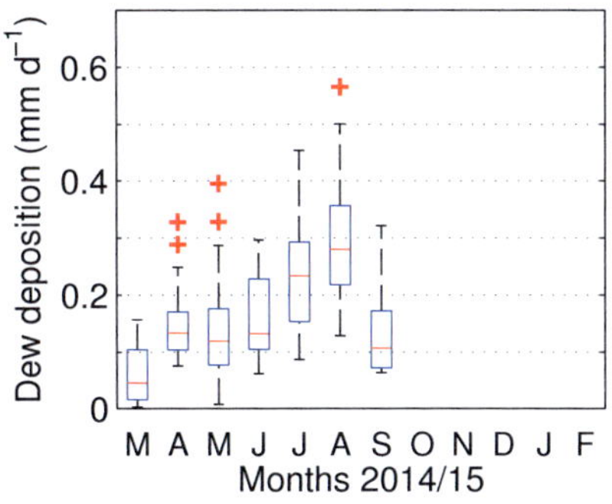

Figure 6.7: Boxplot of daily dew deposition over bare soil at EBS1. Red lines mark medians, the edges of the boxes are the 25th and 75th percentiles, the whiskers extend to the most extreme data points not considered outliers, and outliers are plotted individually as red crosses.

6.5.3 Evaluation of indirect Evaporation Formulas

As evaporation from an open water surface is difficult to measure, it is often estimated by using evaporation formulas. These evaporation formulas are based on distinct indirect approaches and therefore need different input variables to calculate evaporation. With the comprehensive data set of the long-term monitoring network it was possible to apply several of these evaporation formulas and evaluate their applicability for the Dead Sea. Four commonly used approaches to calculate evaporation were applied and compared with the measurements of evaporation at the water surface. The approaches are the Dalton type equation, the energy balance approach, the Penman equation and the Priestley-Taylor equation (Sec. 2.3.2). The measurements have a time resolution of 30 min, but in most studies the formulas are used on a monthly basis and, except of the Dalton type equations, averaging intervals of at least seven days are recommended (Sec. 2.3.2). Therefore, the applicability of the formulas for different averaging intervals was tested, namely 30 min, 1 d, 7 d and 28 d. The so calculated evaporation amounts were compared to the measurements in terms of correlation coefficient, slope and offset of the regression line, mean difference and standard

deviation, and also the intra-annual variation of the differences between estimates and measurements.
As a default version (V0) for the Dalton type approach, the equation after Brutsaert (1982) was used, assuming near neutral conditions (Eq. 2.26 with Eq. 2.27). In the default versions of the other three evaporation formulas the ground heat flux, the net advected heat flux, and also the heat storage term were neglected, as this is a common approach in many studies. To investigate the influence of the performed simplification in the default versions, several sensitivity studies were performed with the equations (Tab. 6.7). The first sensitivity study, afterwards referred to as V1, considers atmospheric stability using Eq. 2.28 and was tested for the Dalton type equation and the Penman equation (Tab. 6.7). Neglecting the heat storage term in the energy balance, Penman, and Priestley-Taylor approach is normally only valid for long time periods, such as 1 yr. As in this study only shorter averaging intervals are investigated, the second sensitivity study (V2) considers the heat storage term using a hysteresis model (Eq. 2.33). Another approach to account for the heat storage term is based on the assumption that the heat storage term is directly proportional to the net radiation and that the missing heat storage term in the default version is responsible for the deviations of the estimated evaporation amounts from the measured amounts. Therefore, an equation for the heat storage term is derived in the form $\Delta Q = x \cdot Rn$, which equals the difference between the estimated evaporation amounts calculated with the default runs and the measured amounts. With this retrieved heat storage term, sensitivity run V3 was performed for the three equations considering the energy balance. For the Penman equation, the only equation which incorporates both the wind function an the energy balance, additional sensitivity studies were performed. In version V4 the surface temperature was removed from the equation, as it was not directly measured, and is therefore another factor of uncertainty. This was implemented by using the parameters after Kohler and Parmele (1967) (Eq. 2.41 and 2.42).

Table 6.7: Overview of the sensitivity studies performed for the evaporation formulas.

Version	Explanation	Dalton	Energy	Priestley-Taylor	Penman
0	Default	X	X	X	X
1	Atmospheric Stability	X	–	–	X
2	Heat storage term with hysteresis model	–	X	X	X
3	Heat storage term derived as a fraction of *Rn* from V0	–	X	X	X
4	Removal of T_s	–	–	–	X
5	Removal of T_s and heat storage term with hysteresis model	–	–	–	X
6	Removal of T_s and heat storage term derived as a fraction of *Rn* from V4	–	–	–	X

The next sensitivity study, V5, combines V2 and V4 incorporating the heat storage term with the hysteresis model and removing the surface temperature using the parameters after Kohler and Parmele (1967). The last sensitivity study (V6) was performed by combining the removal of the surface temperature with a heat storage term derived as a fraction of the net radiation, which equals the differences between the estimated evaporation amounts calculated with V4 and the measured amounts.

Dalton Type Formula

The Dalton type equations are based on the principle of mass transfer at the water surface and describe evaporation as a diffusive flux along the vapour pressure gradient, which also depends on the wind speed. Using the default version of the Dalton type equation after Brutsaert (1982) resulted in high regression coefficient, which increased with longer averaging intervals. The regression coefficients ranged from 0.85 for 30 min averages up to 0.99 for 28 d averages (Tab. 6.8, V0). The slopes of the regression lines were 1.26 for the 30 min averages and around 1.1 for time averages longer than or equal one day, indicating a slight overestimation of high evaporation amounts. The offsets of the

regression lines were small and ranged from -0.01 to -0.34 mm d^{-1}. The good agreement of the estimated and measured evaporation amounts can also be seen in Fig. 6.8 (a,V0) showing the correlation of the estimated amounts calculated from daily averages with the measured values. The mean differences were close to zero for all averaging intervals longer or equal one day, and standard deviation ranged between 0.54 for a 1 d averaging interval and 0.24 mm d^{-1} for 28 d (Fig. 6.9 a). As the mean difference does not show the intra-annual performance of the equation, the differences between the evaporation amounts calculated from the 28 d averages and the measurements are shown in Fig. 6.10 (a). For the default version of the Dalton type equation (Fig. 6.10 a) the differences are small and equally distributed over the year, such that a seasonal bias can be neglected. Accumulating the estimated evaporation amounts and comparing it to the annual evaporation amount measured at EBS2 yields a very good agreement.
The incorporation of the atmospheric stability in the calculations (V1) did not improve the performance of the model. Correlation coefficients were the same as for V0, the offset only decreased slightly for larger time intervals but at the same time the slope increased marginally (Tab. 6.8, V1). No strong deviation from the mean differences or standard deviations from V0 could be found either (Fig. 6.9 a), but a slight overestimation of the annual evaporation amount by 4 % could be observed.

Table 6.8: Slope and offset of the regression lines between the evaporation estimates calculated with the Dalton type formulas and the evaporation measurements and the corresponding correlation coefficient *R*, for averaging periods of 1, 7, 14 and 28 days.

	Slope					Offset					R				
	30 min	1 d	7 d	14 d	28 d	30 min	1 d	7 d	14 d	28 d	30 min	1 d	7 d	14 d	28 d
V0	1.26	1.13	1.08	1.10	1.12	-0.01	-0.33	-0.24	-0.29	-0.34	0.85	0.94	0.94	0.98	0.99
V1	1.27	1.16	1.10	1.13	1.14	-0.01	-0.30	-0.17	-0.23	-0.26	0.85	0.94	0.94	0.98	0.99

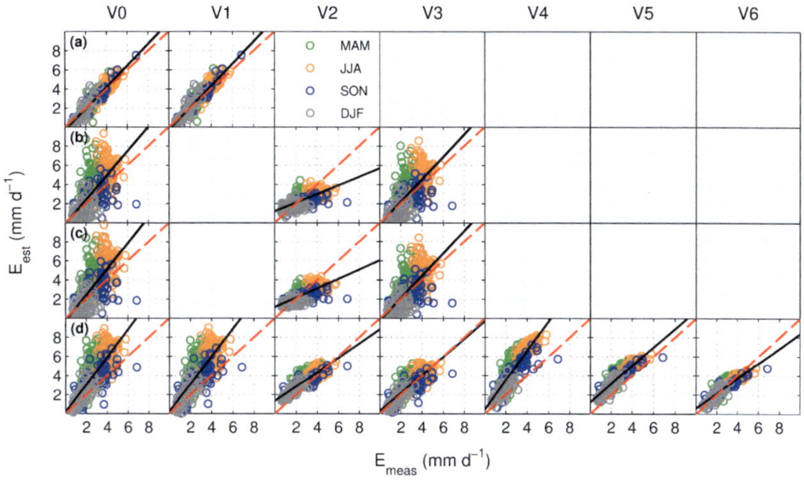

Figure 6.8: Correlation between estimated and measured daily evaporation amounts for (a) the Dalton type equation, (b) the energy balance approach, (c) the Priestley-Taylor equation and (d) the Penman equation and their sensitivity studies (Tab. 6.7) calculated from 1 d averages. The regression line is shown in black and the 1:1 line as dashed red line.

Energy Balance Approach

The energy balance approach is based on the principle to calculate evaporation as the residuum of the energy balance of the water surface by using the Bowen ratio. Thus, only surface water temperature, air temperature and the non-turbulent energy fluxes, i.e. net radiation, ground heat flux, net advected heat flux and the heat storage term, have to be known. Especially the assessment of the heat storage term is problematic, as the lake area, an average lake temperature and the lake volume at the beginning and the end of the time interval are needed. Suggested methods to estimate the heat storage term, like the hysteresis model (Eq. 2.33), which only depends on net radiation, are however, limited in their applicability to longer time intervals. This results in the energy balance approach being not applicable for time averages smaller than one day.

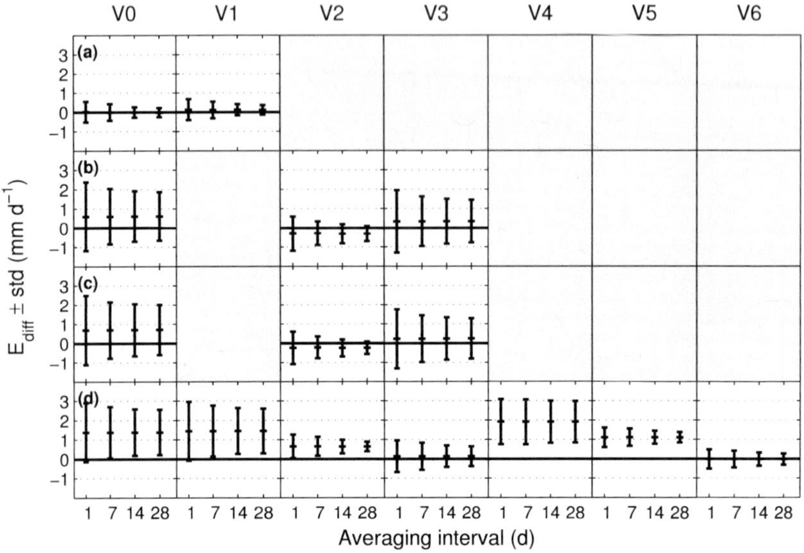

Figure 6.9: Mean differences between the estimated daily evaporation amounts and the measured evaporation amounts for (a) the Dalton type equation, (b) the energy balance approach, (c) the Priestley-Taylor equation and (d) the Penman equation, and their sensitivity studies for an averaging interval of 1,7,14, and 28 days. For further explanation of the sensitivity studies see Tab. 6.7.

Table 6.9: Slope and offset of the regression lines between the evaporation estimates calculated using the energy balance approaches and the evaporation measurements and the corresponding correlation coefficient, *R*, for averaging periods of 1, 7, 14, and 28 days.

	Slope (-)				Offset (mm d^{-1})				R (-)			
	1 d	7 d	14 d	28 d	1 d	7 d	14 d	28 d	1 d	7 d	14 d	28 d
V0	1.27	1.51	1.63	1.72	-0.13	-0.78	-1.11	-1.35	0.67	0.78	0.84	0.87
V2	0.45	0.57	0.63	0.67	1.21	0.89	0.70	0.59	0.69	0.83	0.90	0.96
V3	1.17	1.39	1.50	1.58	-0.12	-0.72	-1.02	-1.24	0.67	0.78	0.84	0.87

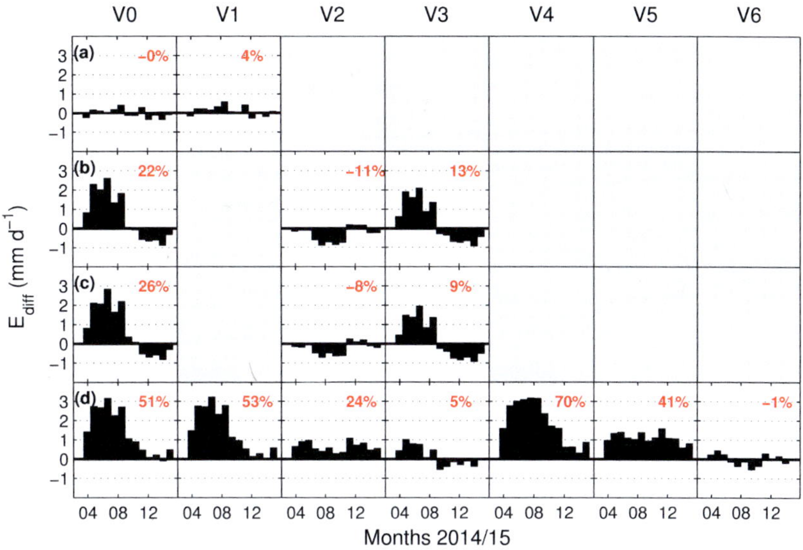

Figure 6.10: Differences between the estimated daily evaporation amounts calculated from the 28 d time averages and the measured evaporation amounts for (a) the Dalton type equation, (b) the energy balance approach, (c) the Penman equation, (d) the Priestley-Taylor equation and their sensitivity studies. The red numbers show the deviation of the accumulated annual evaporation estimate from the measured evaporation amount.

So it was not surprising that neither of the energy balance model versions resulted in a correlation of the estimated and measured values for a time average of 30 min, 3, 6 or 12 h and therefore the results will not be discussed further. For the default version (V0) of the energy balance approach (Eq. 2.32) the ground heat flux, the advected heat flux and the heat storage term were neglected. The results can be ranked after the correlation coefficients for the different averaging intervals. The best agreement was achieved with an time interval of 28 d, leading to a correlation coefficient of 0.87. It decreased with shorter time intervals to only 0.67 for 1 d averages (Tab. 6.9). However, when looking at the slope of the regression line and the offset, it was worst for the 28 d average, with 1.72 for the slope and -1.35 mm d^{-1} as an offset, and best for 1 d, with 1.27 as the slope and

-0.13 mm d^{-1} as an offset (Tab. 6.9). The high slope values and negative offsets implied an overestimation of large evaporation values and an underestimation of small values for all averaging intervals. This is also supported by the differences between the evaporation estimates calculated from the 28 d time averages and the measurements (Fig. 6.10 b). A clear seasonal bias can be identified with an overestimation of the high evaporation values from April until September of up to 3 mm d^{-1} and an underestimation from October until February of up to 1 mm d^{-1}. Similar findings are shown for the evaporation estimates calculated from the 1 d time averages in Fig. 6.8 (b, V0). Evaporation in spring and summer was clearly overestimated and values in winter are underestimated. Compared to the measured values, this is an overestimation by about 50 % and an underestimation of the measured values of up to 100 % (Fig. 6.6 b). This seasonal bias was also observed in other energy balance studies, e.g. Rosenberry et al. (2007) or Winter et al. (1995), and probably results from neglecting the heat storage term in the equation. The mean differences and their standard deviations range from 0.6 ± 1.78 mm d^{-1} for 1 d averages to 0.6 ± 1.28 mm d^{-1} for 28 d time averages (Fig. 6.9 b) and resulted in an overestimation of the annual evaporation amount by 22 %.

To improve the model, the aforementioned sensitivity runs were performed. The hysteresis model (Eq. 2.33) was applied to the data set for the different time averaging intervals longer than or equal 1 d. The constants a, b, and c from Eq. 2.33 were very similar for the different averaging intervals. Therefore, it was decided to use the averages of the constants over the time intervals for calculating the heat storage term of the Dead Sea with

$$\Delta Q = -53.64 + 0.69 \cdot Rn + 0.26 \cdot \frac{\mathrm{d}Rn}{\mathrm{d}t}, \tag{6.6}$$

with Rn and ΔQ in W m^{-2}. Introducing the heat storage term in the model (V2) improved the regression coefficient compared to the default run (Tab. 6.9), and strongly reduced the mean difference and its standard deviation (Fig. 6.9 b). It worked best for the 28 d time average leading to a correlation coefficient of 0.96

and a mean difference of only $-0.3 \pm 0.4\,\mathrm{mm\,d^{-1}}$, which can also be seen in the differences between the estimated daily evaporation amounts calculated from the 28 d time averages and the measured evaporation amounts. The deviations between the estimated and measured values could be reduced and the model performed well between November and April, but underestimated evaporation in summer and autumn by about $1\,\mathrm{mm\,d^{-1}}$, resulting in an underestimation of the annual evaporation amount by 11 % (Fig. 6.10 b). However, the underestimation in spring and autumn results in slopes of the regression lines well below unity, as illustrated for the 1 d averages in Fig. 6.8 (b, V2). The slopes were between 0.45 and 0.67, for the different time intervals and the offset was between 1.2 and $0.6\,\mathrm{mm\,d^{-1}}$, respectively (Tab. 6.9, V2).

When the heat storage term was calculated as a fraction of the net radiation from V0, the calculations resulted in a heat storage term of $\Delta Q = 0.08 \cdot Rn$. As the reduction of the net radiation was not very high (only 8 %), the results of the model (V3) improved only slightly compared to V0 regarding slope, mean difference and standard deviation, and the correlation coefficients remained the same as for the default run (Tab. 6.9, V3). The differences between daily evaporation calculated with the 28 d time averages and the measurements were reduced with regard to the overestimation in spring and summer compared to V0 but at the same time underestimation in winter increased slightly. The total annual evaporation amount was thus improved and was only 13 % above the measured amount (Fig. 6.10 b) compared to the 22 % of V0.

None of the model versions based on the energy balance approach could correctly reproduce the intra-annual variability of Dead Sea evaporation. V0 and V3 had a seasonal bias with strong overestimation in spring and summer and underestimation in winter. V2 underestimated evaporation in summer and autumn and had a large offset of the regression line, but it could reduce the intra-annual deviations to values around or below $1\,\mathrm{mm\,d^{-1}}$ for the 28 d averages and was closest to the annual evaporation amount although the deviation was still -11 %.

Priestley-Taylor Equation

The Priestley-Taylor equation considers the energy balance and the mass transfer approach, but simplifies the part of the mass transfer to a single constant, so that the same variables are needed as for the energy balance approach. For the same reasons as already discussed for the energy balance approach, the Priestley-Taylor equation was not applicable for 30 min time intervals. No correlation of the estimated and the measured values could be found. For the other time averages following results were found. The default model run (V0) of the Priestley-Taylor approach had correlation coefficients of 0.69 for the 1 d average and 0.89 for the 28 d average (Tab. 6.10). The slopes of the regression lines ranged from 1.35 for 1 d averages to 1.84 for 28 d averages and the offset varied from -0.28 mm d^{-1} to -1.58 mm d^{-1}, respectively. The mean difference was about 0.7 mm d^{-1} and the standard deviation varied between 1.81 and 1.3 mm d^{-1} with increasing averaging interval (Fig. 6.9 c). The differences between the estimated and the measured evaporation values for the 28 d time interval showed a clear seasonal bias with an overestimation of up to 3 mm d^{-1} in spring and summer and an underestimation in autumn and winter of about 0.8 mm d^{-1} (Fig. 6.10 c), similar to the results of the energy balance approach, resulting in an overestimation of the annual amount by 26 %.
The same sensitivity runs as for the energy balance approach were performed. The use of Eq. 6.6 to consider the heat storage term of the water led to better results for time intervals of 7 d or larger, with slopes of the regression line between 0.62 for 7 d averages and 0.74 for 28 d averages and smaller offsets from 0.81 to 0.47 mm d^{-1}, respectively (Tab. 6.10, V2). The mean differences was strongly reduced to -0.23 mm d^{-1} for all time averages, and the standard deviations ranged from 0.32 to 0.84 mm d^{-1} (Fig. 6.9 c). The differences between the daily evaporation calculated from the 28 d averages and the measurements showed an underestimation from June until October and quite good results for the other months, resulting in an underestimation of the annual amount by 8 % (Fig. 6.10 c). However, for 1 d results tailed off. The slope of the regression line

was only 0.5 and the offset was 1.17 mm d^{-1} (Fig. 6.8 c, V2), indicating a strong bias in the results.

Similarly as for the energy balance approach, the heat storage term was calculated as a fraction of the net radiation, which equalled the difference between the estimated evaporation amounts from V0 and the measured amounts. The result was $\Delta Q = 0.17 \cdot Rn$. In the Priestley-Taylor equation the net radiation is already multiplied by an empirically determined coefficient $c_{PT} = 1.26$ to account for the aerodynamic influence. Combining the aerodynamic influence with the heat storage term resulted in a new Priestley-Taylor constant of $c_{PT} = 1.09$ used in model version V3. However, this did only improve the model performance for a time average of 1 d. The slope was better and the offset was smaller compared to V0 and V2 (Fig. 6.8 c, V3), but for the 7 to 28 d time averages the results were worse compared to V2, with slopes between 1.39 and 1.59 and large offsets. The differences between the daily evaporation amounts calculated from the 28 d time averages and the measurements showed a seasonal bias with overestimation up to 2 mm d^{-1} in spring and summer and an underestimation of 1 mm d^{-1} in autumn and winter (Fig. 6.10 d). The total annual deviation was comparable to version V2 with -9 %.

When using the Priestley-Taylor equation with time averages longer than one day, model version V2 gives the best results of estimated evaporation, even though it generally underestimated high evaporation values and overestimates small values. For 1 d averages model version V3 yielded the best results, but the spread of the results was quite high.

Penman Equation

The Penman approach combines the mass transfer and the energy balance equation. Besides the variables already needed for the energy balance or the Prietley-Taylor equation, wind speed and an empirical wind function are required for this equation (Tab. 2.1). For the Penman approach the results of the default model version and its sensitivity runs are shown in Tab. 6.11.

Table 6.10: Slope and offset of the regression lines between the evaporation estimates calculated with the Priestley-Taylor equation and the evaporation measurements and the corresponding correlation coefficient *R*, for averaging periods of 1, 7, 14 and 28 days.

	Slope				Offset				R			
	1 d	7 d	14 d	28 d	1 d	7 d	14 d	28 d	1 d	7 d	14 d	28 d
V0	1.35	1.61	1.74	1.84	-0.28	-0.98	-1.33	-1.58	0.69	0.80	0.86	0.89
V2	0.49	0.62	0.70	0.74	1.17	0.81	0.59	0.47	0.73	0.87	0.93	0.98
V3	1.17	1.39	1.51	1.59	-0.24	-0.85	-1.15	-1.37	0.69	0.80	0.86	0.89

The Penman approach also considers the heat storage term of the lake and thus does not give reasonable results for time averages in the sub-daily range. The correlation coefficient for the default model for 30 min was 0.16 and was similar for the sensitivity runs. For longer time intervals the Penman approach was applicable, and the correlation coefficient for the default model run was 0.78 for time averages of 1 d and increased with increasing averaging interval to 0.91 for 28 d (Tab. 6.11, V0). Despite the good correlation coefficients, the slope of the regression line was 1.44 and the offset was 0.19 mm d^{-1} for 1 d and even worse for longer averaging intervals (Tab. 6.11, V0). This indicates that the default model generally overestimated evaporation. For all time intervals mean difference and standard deviation were high with values around 1.38$\pm$1.52 mm d^{-1} (Fig. 6.9 d). The differences between the estimated evaporation amounts calculated from the 28 d averages and the measured amounts show the overestimation clearly (Fig. 6.10 d). Evaporation was overestimated by up to 3 mm d^{-1} in spring and summer and by about 1 mm d^{-1} in autumn, which is an overestimation of nearly 100% on the measured values in some months (Fig. 6.6 b). Only in winter the values were close to the observed amounts. Overall it resulted in an overestimation of the annual evaporation amount by 51 %. Sensitivity runs were performed to reduce the overestimation of the evaporation. In the first sensitivity run atmospheric stability was taken into account by using Eq.2.28 but did not improve the model at all. The correlation coefficients were the same as for the default run and also slope and offset were comparable (Tab. 6.11, V1 and Fig. 6.8 d).

The second variation (V2) considered the heat storage term calculated with the hysteresis model (Eq. 6.6). This did not only improve the correlation coefficient but also reduced the mean difference to 0.65 mm d^{-1} and the standard deviation to 0.61 mm d^{-1} for 1 d averages and even 0.25 mm d^{-1} for 28 d averages (Fig. 6.9 d). This means that the spread in the estimated values was strongly reduced, which is illustrated for the values calculated from 1 d time averages in Fig. 6.8 (d, V2). The slope of the correlation line between estimates and measurements was 0.75 for 1 d and increased to 0.89 for 28 d average, with an offset of 1.34 and 0.94 mm d^{-1}, respectively (Tab. 6.11, V2), meaning that small values were still highly overestimated. This is also apparent in the intra-annual deviations of the estimated evaporation amounts from the measurements. The deviations were reduced to values around 1 mm d^{-1} (Fig. 6.10 d) and are thus lower in summer and autumn compared to V0 and V2, but higher in spring and winter. The annual overestimation could be reduced to 24 %.

When calculating the heat storage term as a fraction of the net radiation from the results of the Penman model run V0, a heat storage term of $\Delta Q = 0.46 \cdot Rn$ was obtained. This is considerably higher than the results from the energy balance approach, where it was only 8 % of the net radiation. It is the consequence of the strong overestimation of the evaporation in the Penman model V0. The model run which considered the so calculated heat storage term (V3) achieved slopes of the regression lines close to one and small offsets between 0.29 mm d^{-1} for 1 d averages (Fig. 6.8 d, V3) and -0.11 mm d^{-1} for 28 d averages (Tab. 6.11, V3). The correlation coefficients were also high, with a correlation of 0.82 for 1 d averages and 0.92 for 28 d averages. Also the mean differences were small, with 0.13 mm d^{-1} for all time averages and standard deviations of the mean difference of 0.51 to 0.82 mm d^{-1} (Fig. 6.10 d). This model version overestimated evaporation in spring and summer by about 1.0 mm d^{-1} and underestimated it in autumn and winter by about 0.5 mm d^{-1} or less (Fig. 6.10 d). Compared to the previously mentioned model versions, this model has a small seasonal bias, which is most likely caused by neglecting the hysteresis of the heat storage, as the heat storage term is estimated as a mere fraction of the net radiation. The annual deviation from the measured amount was 5 %.

The removal of the surface temperature by using the parameters after Kohler and Parmele (1967) did improve the correlation coefficients of the model runs to values between 0.89 and 0.96 for the corresponding time averaging intervals, but the slope worsened to values between 1.54 for 1 d averages and 1.87 for 28 d averages (Tab. 6.11, V4). The mean difference increased to $1.91 \pm 1.16\,\mathrm{mm\,d^{-1}}$, also indicating a strong overestimation (Fig. 6.9 d). The overestimation can also be seen in Fig. 6.10 (d) where deviations of up to $3\,\mathrm{mm\,d^{-1}}$ were calculated for April until August, resulting in an overall overestimation of the annual evaporation amount by 70 %.

When using the hysteresis model for the lake heat storage term and the parameters after Kohler and Parmele (1967), correlation coefficients above 0.92 were achieved and the slope was close to unity for all time averages (Tab. 6.11, V5). However, the offset varied between 0.93 and $1.22\,\mathrm{mm\,d^{-1}}$, meaning that evaporation was generally overestimated by about $1\,\mathrm{mm\,d^{-1}}$ (Fig. 6.10 d, 6.8 d, V5), which was also represented by a mean difference of $1.1\,\mathrm{mm\,d^{-1}}$ (Fig. 6.9 d).

The last model version, V6, used the parameters after Kohler and Parmele (1967), and the heat storage term was calculated as a fraction of the net radiation, which corresponded to the deviation of the evaporation amounts calculated with V4 and the measured amounts. This resulted in a very high heat storage term of $\Delta Q = 0.77 \cdot Rn$. However, this version achieved good correlation coefficients, with values between 0.92 and 0.97, and had very low mean differences of -$0.02\,\mathrm{mm\,d^{-1}}$ and standard deviations of only 0.29 to $0.5\,\mathrm{mm\,d^{-1}}$ (Fig. 6.9 d), but the slope and the offset of the regression line were not as good as e.g. in V3 (Tab. 6.11). The slope ranged from 0.78 to 0.84 for increasing averaging intervals and the offset ranged from 0.59 to $0.42\,\mathrm{mm\,d^{-1}}$, respectively, meaning that high evaporation values were generally underestimated and small values overestimated. Differences between the daily evaporation amounts calculated from the 28 d averages and the measurements were all below $0.5\,\mathrm{mm\,d^{-1}}$ (Fig. 6.10 d), resulting in a very good agreement of the annual evaporation amount with the measured one.

The default version of the Penman equation (V0), the consideration of atmospheric stability (V1) and the removal of the surface temperature from the equation (V4) did not satisfyingly estimate evaporation. The slopes of the regression lines were all too high in those models, implying an overestimation of large evaporation amounts. Using the hysteresis model for estimating the heat storage term (V2) or using the hysteresis model and removing the surface temperature (V5) did improve the slope but the offsets were too high, leading to a stronger overestimation of small amounts and a weaker overestimation of high amounts, resulting in a general overestimation of the evaporation regardless of the time of the year. Best results were achieved with the equation especially derived for the Dead Sea, where the heat storage term was only considered as a fraction of the net radiation (V3 or V6), making this equation applicable for the Dead Sea.

Table 6.11: Correlation coefficient *R*, slope, and offset of the regression line for evaporation calculated with the Penman equation versus measurements, for averaging periods of 1, 7, 14 and 28 days.

	Slope				Offset				R			
	1 d	7 d	14 d	28 d	1 d	7 d	14 d	28 d	1 d	7 d	14 d	28 d
V0	1.44	1.58	1.69	1.76	0.19	-0.20	-0.49	-0.69	0.78	0.83	0.88	0.91
V1	1.44	1.57	1.68	1.76	0.24	-0.12	-0.41	-0.61	0.78	0.83	0.88	0.91
V2	0.75	0.80	0.85	0.89	1.34	1.21	1.04	0.94	0.87	0.89	0.94	0.97
V3	0.94	0.99	1.05	1.09	0.29	0.16	0.00	-0.11	0.82	0.84	0.90	0.92
V4	1.54	1.73	1.80	1.87	0.45	-0.09	-0.28	-0.48	0.89	0.91	0.94	0.96
V5	0.96	1.01	1.04	1.06	1.22	1.10	1.00	0.93	0.92	0.93	0.96	0.97
V6	0.78	0.80	0.81	0.84	0.59	0.54	0.49	0.42	0.92	0.92	0.95	0.97

6.6 Discussion

With a comprehensive data set of energy balance measurements from the typical land surfaces in the Dead Sea valley, bare soil, water, vegetation, an in-depth analysis of the land surface - atmosphere interaction was possible. It was shown that it is possible to set up regression models for the sensible and latent heat flux from the water surface with data exclusively measured at the shoreline. The models are independent of the data used to build the model and are robust enough to calculate the turbulent fluxes from the water surface with a time resolution of 30 min (Tab. 6.4, 6.3). The latent heat flux depends on wind speed and vapour pressure deficit, which generally agrees with studies performed for other lakes, e.g. Blanken et al. (2000); Mammarella et al. (2015); Nordbo et al. (2011). However, in these studies the best correlation of the latent heat flux was achieved with the product of wind speed and vapour pressure deficit, whereas for the Dead Sea best correlation was found with the sum of wind speed and vapour pressure deficit. The sensible heat flux depends on wind speed, the temperature deficit of the surface water temperature and the air temperature, and the product of the two aforementioned variables, which also coincides with the aforementioned studies. The surface water temperature can be calculated using the Monin-Obukhov similarity theory, which eliminates the need of direct measurements of the surface water temperature (Sec. 4.3.2).

Furthermore, the results indicate that over bare soil about 60 to 80 % of the net radiation is transformed into sensible heat flux and only 9 to 17 % into latent heat flux due to the limitation in soil moisture (Tab.6.5). Over water it is the opposite: sensible heat flux is very small and latent heat flux dominates. From spring until autumn only 5 to 10 % of the net radiation is transformed into sensible heat flux, but 43 to 73 % are transformed into latent heat flux (Tab.6.5). The rest of the energy leads to a heating of the water body, due to the high heat capacity of the water. In winter, heat is released from the water body, leading to a higher sensible heat flux in winter. At EBS3 both fluxes have to be considered as the transpiration of plants increases the measured evapotranspiration considerably.

The differences in the energy balance partitioning at the different surfaces leads to strong horizontal gradients of sensible heat flux between the water surface and the land surface during daytime, resulting in a stronger heating of the air mass above the land compared to the air above the lake. During the night the sensible heat flux gradient is reversed, but much weaker (Fig. 6.3).
The median daily evaporation amounts from the water and from the vegetated areas follow a clear annual cycle with median daily evaporation amounts of 4.4 mm d^{-1} in summer (Fig. 6.6). Additionally, there is nearly no precipitation in summer, which leads to the conclusion, that the loss of water from the hydrological system is highest during the summer months and that the lake level declines strongest in summer. Given that more than half of the valley floor is covered with water, the consistently high latent heat flux from the water surface results in a considerable water vapour flux from the water body to the atmosphere. In summer, about $2.58 \cdot 10^9$ l of water are released from the water surface to the atmospheric boundary layer per day. Although highest median evaporation amounts at the water surface are measured in summer, the absolute maximum of daily evaporation during the investigation period was observed in November (Fig. 6.6). Evaporation amounts of up to 6.86 mm d^{-1} were measured during a Red Sea Trough (RST) event. Thus, it can be concluded that extreme evaporation events are triggered by synoptic systems which influence the atmospheric conditions in such a way that evaporation amounts even exceed summer evaporation rates. This means constantly high wind velocities and the advection of dry air which causes high vapour pressure deficits, can cause extreme evaporation events. This result is particularly important for the Dead Sea evaporation as Alpert et al. (2004) found that the frequency of such RST systems nearly doubled since the 1960s from 50 to 100 days per year and that the whole Dead Sea region is highly sensitive to climate change (Alpert et al., 1997; Smiatek et al., 2011).
The comparison with common evaporation formulas showed that only the Dalton formula was able to produce unbiased and reliable evaporation estimates for sub-daily time intervals but also for averaging periods longer than 1 d. The standard deviation of the estimates was 0.54 mm d^{-1} or less and the annual evaporation

amount was estimated correctly. All other commonly used approaches had large difficulties reproducing the intra-annual variability of evaporation, especially when the heat storage term of the lake was not considered in the calculations. The results showed an overestimation of daily evaporation amounts in summer by over 50 % and an underestimation in winter by nearly 100 %. Hence, it can be concluded that the heat storage term can not be neglected for the calculation of the Dead Sea evaporation, when time intervals of 28 d or less are analysed. The usage of a hysteresis model to account for the heat storage term did improve the results but they were still not as good as the Dalton type formula. The results of the energy balance and Priestley-Taylor approach showed a seasonal bias with an underestimation of daily evaporation amounts in summer and an overestimation in winter. This resulted in a deviation of the estimated annual evaporation amount from the measured amount by -11 % for the energy balance approach and -8 % for the Priestley-Taylor approach, respectively. Results comparable to the Dalton type formula could only be achieved using the Penman equation with the adjusted heat storage term either obtained from the default version or from the version were the surface water temperature was eliminated from the equation (V4). Using $\Delta Q = 0.46 \cdot Rn$ resulted in a weak seasonal bias with overestimation in spring and weak underestimation in autumn, and the annual amount was only overestimated by 5 %. Using the parameters after Kohler and Parmele (1967) and $\Delta Q = 0.77 \cdot Rn$ further reduced the bias and estimated the annual amount correctly. As the Penman and the Dalton type equation both consider wind speed and vapour pressure deficit as governing factors of the evaporation process, this result supports the established regression model for the latent heat flux, which also depends on wind speed and vapour pressure deficit over the lake. It can be concluded that measurements of the heat storage term are inevitable for the correct application of the energy balance and the Priestley-Taylor method at the Dead Sea. Whereas for the Penman equation, the heat storage term can be estimated by using an empirically gained function, to correctly calculate evaporation. The Dalton type equation can be applied without any corrections and should therefore be favoured.

7 Summary of Achievements

The Dead Sea is a unique place on earth and it faces severe environmental challenges concerning the atmosphere, hydrosphere and lithosphere. The over-exploitation of water resources leads to an extreme lake level decline. This has severe implications, such as the shifting of the fresh/saline groundwater interface, drying up of the lake, and sinkhole formation. Moreover, it leads to climatic changes and the dry lake bed serves as a source for dust emissions. Taking the meteorological perspective, two key features are relevant. Firstly, wind systems, which are essential for the local climate in the valley. They determine the atmospheric conditions, influence air quality and visibility. Secondly, the energy balance of the Earth's surface. It drives local wind systems and it links the different spheres. Changes in the energy balance can affect the atmosphere, as it is responsible for the heat exchange and moisture supply. It can affect the hydrosphere, as evaporation is an important component of the Dead Sea water budget determining the lake level decline, and the lithosphere, determining soil temperature and soil moisture.

Neither comprehensive three-dimensional measurements of the wind systems to describe and understand their specific trigger mechanisms, their three-dimensional structure and their influence on the valley's atmosphere, nor long-term measurements of the energy balance of the main surfaces in the valley have been performed in the Dead Sea valley so far. However, these measurements were recently realised in the framework of DESERVE. Long-term measurements of the near-surface atmospheric conditions in the valley and on the slopes were performed, together with eddy-covariance measurements over the three main surfaces in the valley, over bare soil, water, and vegetation. The long-term measurements were complemented by two SOPs investigating the three-dimensional conditions of the atmosphere with sophisticated instruments, such as lidar systems, radiosondes, or GPS. Finally, laboratory measurements were performed to obtain the water characteristics necessary for the analysis. With the composed comprehensive database the scientific objectives of this thesis can be answered.

(1) Intra-annual variation and frequency of recurring wind systems in the Dead Sea valley, their characteristics, three-dimensional structure, and trigger mechanisms.

Several distinct wind systems could be identified in the Dead Sea valley. The long-term measurements revealed that during daytime upslope winds and a lake breeze, in the evening DWFs, and during nighttime northerly along-valley flows prevail.

The upslope winds are initiated by the differential heating along the slopes, and the lake breeze is triggered by the differential heating between the land surface and the water surface during daytime. The results indicate that the lake breeze suppresses expected up-valley winds, but superimposes with the upslope winds. It causes a delay of the diurnal temperature increase in the valley and it leads to steady, moderate wind velocities. The height of the lake breeze circulation is determined by the temperature inversion at the top of the valley CBL, which can reach above ridge height. The long-term measurements show that during spring, summer, and autumn a lake breeze was observed on up to 70 % of the days. In winter, the lake breeze was also observable on 45 % of the days, even though a superposition of the large-scale forcing caused a shift of the wind direction to south-east in about half of cases. The intra-annual variability of the lake breeze occurrence, with higher frequencies in summer and lower frequencies in winter, reflects the stronger synoptic influence over whole Israel in winter.

The strong westerly DWFs occur in the evening, especially in summer, but also in spring and autumn. The results confirm that the DWFs are triggered by the temperature difference between the air masses in the valley and around ridge height, leading to a density current which accelerates on its way down into the valley. This temperature difference is caused either by the prolonged warming of the air in the valley and radiative cooling at the ridge or through additional advection of cooler maritime air masses towards the ridge by the Mediterranean sea breeze front. A detailed analysis of the characteristics of the DWFs was performed using lidar data. An algorithm was developed to automatically detect the height of the DWFs, as well as the mean and maximum wind speed and its duration. Three different types of DWFs were identified: short, medium, and long lasting events. Short DWFs have moderate mean and maximum wind velocities of about 3 to $6\,\mathrm{m\,s^{-1}}$ and a layer height of about 600 m. They descend into the valley but are not able to penetrate far into the valley, only about 3 to 5 km, and are stopped by nocturnal northerly along-valley flows. Medium DWFs last about 4 to 5 h and reach mean maximum wind speeds of over $10\,\mathrm{m\,s^{-1}}$. They influence the whole valley atmosphere up to the ridge and more than 13 km into the valley by causing a total exchange of the valley air mass. The results of the COSMO-EU analysis indicate that these events are most likely intensified by a secondary trigger mechanism upstream which leads to a transition of a subcritical to a supercritical flow regime. The last type of DWFs are long lasting moderate events with a very shallow height of about 250 to 400 m. Their mean wind speed is moderate and comparable to the short events. Although these three types were identified, a transition of a strong DWF to a longer lasting moderate DWF is also possible.

The nocturnal northerly along-valley flows are driven by the along-valley temperature difference between the northern part and the southern part. The cooling in the northern part was more effective than in the south, most likely caused by a combination of the typical cooling mechanisms in the valley, e.g. the TAF, and the fact that the southern evaporation ponds are shallower and thus much warmer than the Dead Sea. The warmer water results in a stronger heat input from the ponds to the atmosphere in the evening and a delayed cooling of the valley atmosphere in the south. The northerly along-valley flows have a vertical extent of about 600 m, which is about half of the valley's depth. Mean horizontal wind speeds are moderate, but they have a jet like structure with the wind speed maximum at about 200 to 300 m above the ground. The northerly along-valley flows are only observed in spring, summer, and autumn.

(2) Energy balance partitioning at the main surfaces in the Dead Sea valley, i.e. bare soil, water, and vegetation, the differences and the influence on the wind field and on the local climatic conditions.

Over bare soil the sensible heat flux greatly exceeds the latent heat flux, whereas over water it is the opposite: the sensible heat flux is small and the latent heat flux is large. The strong horizontal gradient of the sensible heat flux between water and land surface results in a differential heating of the air over the water and over land, which causes a steady and persistent lake breeze during daytime. Besides, the different energy balance partitioning at the different surfaces has further impacts on the atmosphere and the underlying soil. Over bare soil the limitation of available soil moisture leads to a daily evaporation below 1 mm d^{-1}, and the net radiation is almost entirely transformed into sensible heat and soil heat flux, resulting in high soil temperature and high air temperature. Additionally, the conditions are suitable for nocturnal dew deposition. Dew deposition regularly occurs after sunset and serves as an important water source for plants and microbiotic crusts. Over vegetation the latent heat flux is much higher compared to bare soil because the plants are able to reach deeper groundwater layers with their roots. This leads to ten times higher daily evapotranspiration amounts compared to bare soil. Furthermore, an oasis effect, creating a microclimate over the vegetated area, was observed. Through enhanced wind velocities in the evening, evapotranspiration increases strongly. The necessary energy for the evapotranspiration is provided by a cooling of the air and thus a negative sensible heat flux, and this leads to colder nights in the vegetated areas compared to the unvegetated areas. The difference between the air temperature over bare soil and vegetation is up to 5 °C. Energy balance partitioning at the water surface is determined by the latent heat flux and heat storage. Through the unlimited water supply and the high heat storage capacity of the water, energy is either transformed into latent heat or it heats the water. The evaporation from the water surface serves as an important moisture source for the valley atmosphere, as the valley ABL is mainly decoupled from the large-scale flow and thus from the advection of moisture from other sources. With a current surface area of the lake of 600 km^2 and a median summer evaporation of 4.3 mm d^{-1}, this results in $2.58 \cdot 10^9$ l water, which is released from the water surface to the atmospheric boundary layer per day. Through this high moisture input, the evaporation provides the necessary moisture for the dew deposition over land and haze formation.

(3) The actual evaporation rate of the Dead Sea water and its intra-annual variability.

Evaporation from the water surface has an annual cycle which generally follows the solar cycle, with highest median values of 4.3 mm d^{-1} in July and lowest values in December. However, extreme evaporation events were observed in winter, with the absolute maximum of daily evaporation in the investigation period, which was 6.86 mm d^{-1}, triggered by the large-scale conditions. On the diurnal scale evaporation is not connected to solar radiation. Instead it is clearly governed by the diurnal cycle of the wind systems. During the day only moderate evaproation amounts are reached as the lake breeze has only low wind velocities between 2 and 4 m s^{-1}. In the evening the DWFs cause maximum evaporation rates due to their high wind velocities of over 10 m s^{-1}, and during the night evaporation rates are higher compared to daytime values, resulting from the accelerated wind velocities of the nocturnal along-valley flows. The accumulated amount of

Dead Sea water evaporation measured near Ein Gedi was 994.5±81.2 mm. This result supports early evaporation estimates which were at the bottom of the evaporation estimate range, such as Stanhill (1994) with 1.05 m yr^{-1} or Lensky et al. (2005) with 1.2 m yr^{-1}.

(4) Multiple regression model for the turbulent heat fluxes of the Dead Sea water surface and an investigation of the reliability of commonly used evaporation formulas on different time scales.

A multiple linear regression model for the sensible and latent heat flux was set up and the results show that the turbulent fluxes can not be described by one meteorological variable only. The sensible heat flux depends on wind speed and on the temperature gradient between surface water temperature and air temperature and the latent heat flux depends on wind speed and vapour pressure deficit. The surface water temperature could be calculated from the air temperature using the Monin-Obukhov similarity theory. The robustness of the model was tested using a Monte-Carlo cross validation, and the model yields good results when it is applied to one of the meteorological seasons or to the whole data set. As the measurement site is located at the shoreline, it is possible to calculate the turbulent fluxes with this model for situations where the flux footprint is not located over the water surface and thus achieve a complete picture of the turbulent fluxes under all possible wind conditions.

The comparison with evaporation formulas shows strong deficits of several equations to correctly estimate evaporation on an intra-annual scale. Specifically, the equations which consider the energy balance of the water are difficult to apply. They are not applicable for time intervals below 1 d and generally yield a strong over-estimation of summer evaporation, but an underestimation of the winter evaporation, when the heat storage term is not considered. The differences are up to 100 % on a monthly basis, but for an annual estimate the aforementioned under- and overestimation can compensate each other resulting in an annual amount which is only

20 % above the measured annual amount. The use of a hysteresis model to account for the heat storage term in the equations improved the estimates, but still yielded considerable deviations from the measurements. With an adjusted heat storage term for the Dead Sea, the Penman equation yielded satisfying results but only for daily or longer time intervals. The only equation which was able to produce unbiased and reliable evaporation amounts for sub-daily time intervals, but also on daily, weekly or monthly scales, was the Dalton formula. It yielded good results with small deviations from the measurements. This also supports the established regression model because both, the Dalton formula and the model, describe evaporation as a function of wind speed and vapour pressure deficit.

8 Conclusions

The results achieved in this thesis lead to following conclusions about the conditions in the Dead Sea valley, particularly with regard to the ongoing rapid change of the land to water surface ratio in the valley. It was concluded that, based on the results, a shrinking water surface area leads to a weaker horizontal sensible heat flux gradient in the valley, which results in a weakening of the lake breeze. This will change the climatic conditions, as the attenuating effect of the water on the air temperature will weaken, leading to lower minimum and to higher maximum air temperatures. Such changes will affect living and working conditions in the area, as well as the tourist and health sector. The increasing heat stress will result in outdoor activities becoming physically more strenuous, and will have to be taken into account when planning new constructional projects. A weaker lake breeze could be overcome by a currently suppressed up-valley flow, which would change the wind field in the valley. A southerly wind could transport the emitted air pollutants from the southern industrial area to the north, where the main hotel and resort district as well as the larger villages are located. Furthermore, as a result of a shrinking water surface area, the water input to the atmosphere will be reduced, which can affect the haze formation and can cause a reduction of the dew deposition, as the necessary high moisture in the atmosphere may not be available. On the annual scale, winter evaporation could become more important in the future. It was shown that extreme evaporation events are triggered by synoptic forcing such as a strong Red Sea Trough. If the frequency of these synoptic systems will increase in the future, winter evaporation will increase as well. The amount of sensible heat released to the atmosphere increases as more land surface is exposed due to the shrinking of the

water surface. This intensified sensible heat release can lead to stronger convection and stronger vertical mixing influencing the boundary layer development. In addition, it can further increase the frequency of DWFs, as the necessary temperature difference between the valley air mass and the air mass at ridge height is easier reached with higher temperatures in the valley.
The gained insights into the specific wind conditions led to the conclusion that local radiative cooling and delayed warming in the valley are sufficient to initiate DWFs. These results can help to understand the wind conditions in other parts of the world with complex terrain and weak synoptic influence. Complex terrain, such as mountains, plateaus, and hills, cover about half of the Earth's surface and especially the valleys are often densely populated. Therefore, the detailed understanding and forecast of the local valley wind conditions is imperative. For future investigations of the conditions in the Dead Sea valley the obtained results of this thesis can be used to reliably forecast evaporation, provided that a forecast of wind speed and temperature is available.

Bibliography

Abels, D. J. and J. Kattan-Byron, 1985: Psoriasis treatment at the Dead Sea: a natural selective ultraviolet phototherapy. *J. Am. Acad. Dermatol.*, **12 (4)**, 639–643.

Abels, D. J., T. Rose, and J. E. Bearman, 1995: Treatment of psoriasis at a Dead Sea dermatology clinic. *Int. J. Dermatol.*, **34 (2)**, 134–137.

Abelson, M., Y. Yechieli, O. Crouvi, G. Baer, D. Wachs, A. Bein, and V. Shtivelman, 2006: Evolution of the Dead Sea sinkholes. *Geol. Soc. Spec. Pap.*, **401**, 241–253.

Allwine, K. J., 1993: Atmospheric dispersion and tracer ventilation in a deep mountain valley. *J. Appl. Meteorol.*, **32 (6)**, 1017–1037.

Alpert, P., R. Abramsky, and B. U. Neeman, 1990: The prevailing summer synoptic system in Israel - Subtropical high, not Persian Trough. *Israel J. Earth Sci.*, **39 (2/4)**, 93–102.

Alpert, P., I. Osetinsky, B. Ziv, and H. Shafir, 2004: Semi-objective classification for daily synoptic systems: application to the eastern Mediterranean climate change. *Int. J. Climatol.*, **24 (8)**, 1001–1011.

Alpert, P., H. Shafir, and D. Issahary, 1997: Recent changes in the climate at the Dead Sea - a preliminary study. *Climatic Change*, **37 (3)**, 513–537.

Anderson, D. E., S. B. Verma, and N. J. Rosenberg, 1984: Eddy correlation measurements of CO2, latent heat, and sensible heat fluxes over a crop surface. *Boundary-Layer Meteorol.*, **29 (3)**, 263–272.

Antonelli, M. and R. Rotunno, 2007: Large-eddy simulation of the onset of the sea breeze. *J. Atmos. Sci.*, **64 (12)**, 4445–4457.

Arkin, Y. and A. Gilat, 2000: Dead Sea sinkholes - an ever-developing hazard. *Environ. Geol.*, **39 (7)**, 711–722.

Armi, L. and G. J. Mayr, 2011: The descending stratified flow and internal hydraulic jump in the lee of the Sierras. *J. Appl. Meteor. Climatol.*, **50 (10)**, 1995–2011.

Arritt, R., 1993: Effects of the large-scale flow on characteristic features of the sea breeze. *J. Appl. Meteorol.*, **32**, 116 – 125.

Arya, S. P., 2001: *Introduction to micrometeorology*, International Geophysics Series, Vol. 79. 2d ed., Academic Pr., San Diego, Calif. [u.a.], 354 pp.

Askne, J. and H. Nordius, 1987: Estimation of tropospheric delay for microwaves from surface weather data. *Radio Sci.*, **22 (3)**, 379–386.

Asmar, B. N. and P. Ergenzinger, 1999: Estimation of evaporation from the Dead Sea. *Hydrol. Processes*, **13 (17)**, 2743–2750.

Assouline, S. and Y. Mahrer, 1993: Evaporation from Lake Kinneret: 1. Eddy correlation system measurements and energy budget estimates. *Water Resour. Res.*, **29 (4)**, 901–910.

Atkins, P., 2014: *Physical chemistry*. 10th ed., Oxford University Press, Oxford, 1008 pp.

Atkinson, B., 1981: *Meso-scale atmospheric circulations*. Academic Press, London, 495 pp.

Banta, R. M., C. M. Shun, D. C. Law, W. Brown, R. F. Reinking, R. M. Hardesty, C. J. Senff, W. A. Brewer, M. J. Post, and L. S. Darby, 2013: Observational techniques: Sampling the mountain atmosphere. *Mountain Weather Research and Forecasting: Recent Progress and Current Challenges*, Chow, K. F., F. S. De Wekker, and J. B. Snyder, Eds., Springer Netherlands, Dordrecht, 409–530.

Barthlott, C. and D. J. Kirshbaum, 2013: Sensitivity of deep convection to terrain forcing over Mediterranean islands. *Q. J. R. Meteorol. Soc.*, **139 (676)**, 1762–1779.

Bevis, M., S. Businger, S. Chiswell, T. A. Herring, R. A. Anthes, C. Rocken, and R. H. Ware, 1994: GPS meteorology: Mapping zenith wet delays onto precipitable water. *J. Appl. Meteorol.*, **33 (3)**, 379–386.

Bevis, M., S. Businger, T. A. Herring, C. Rocken, R. A. Anthes, and R. H. Ware, 1992: GPS meteorology: Remote sensing of atmospheric water vapor using the Global Positioning System. *J. Geophys. Res. Atmos*, **97 (D14)**, 15 787–15 801.

Bitan, A., 1974: The wind regime in the north-west section of the Dead-Sea. *Arch. Meteorol. Geophys. Bioklim. Ser B*, **22 (4)**, 313–335.

———, 1976: The influence of the special shape of the Dead-Sea and its environment on the local wind system. *Arch. Meteorol. Geophys. Bioklim. Ser B*, **24 (4)**, 283–301.

Bitan, A. and H. Sa'Aroni, 1992: The horizontal and vertical extension of the Persian Gulf pressure trough. *Int. J. Climatol.*, **12 (7)**, 733–747.

Blanken, P. D., W. R. Rouse, A. D. Culf, C. Spence, L. D. Boudreau, J. N. Jasper, B. Kochtubajda, W. M. Schertzer, P. Marsh, and D. Verseghy, 2000: Eddy covariance measurements of evaporation from Great Slave Lake, Northwest Territories, Canada. *Water Resour. Res.*, **36 (4)**, 1069–1077.

Bolton, D., 1980: The computation of equivalent potential temperature. *Mon. Weather Rev.*, **108 (7)**, 1046–1053.

Brazel, A., H. Fernando, J. Hunt, N. Selover, B. Hedquist, and E. Pardyjak, 2005: Evening transition observations in Phoenix, Arizona. *J. Appl. Meteorol.*, **44 (1)**, 99–112.

Browning, K. and R. Wexler, 1968: The determination of kinematic properties of a wind field using Doppler radar. *J. Appl. Meteorol.*, **7 (1)**, 105–113.

Brutsaert, W., 1982: *Evaporation into the Atmosphere. Theory, History, and Applications*. Reidel, Dordrecht, Netherlands, 299 pp.

Burba, G., 2013: *Eddy covariance method for scientific, industrial, agricultural and regulatory applications: A field book on measuring ecosystem gas exchange and areal emission rates*. LI-Cor Biosciences, Lincoln, Nebraska, USA, 331 pp.

Calder, I. and C. Neal, 1984: Evaporation from saline lakes: a combination equation approach. *Hydrol. Sci. J.*, **29 (1)**, 89–97.

Cline, D. W., 1997: Snow surface energy exchanges and snowmelt at a continental, midlatitude Alpine site. *Water Resour. Res.*, **33 (4)**, 689–701.

Crosman, E. T. and J. D. Horel, 2010: Sea and lake breezes: A review of numerical studies. *Boundary-Layer Meteorol.*, **137 (1)**, 1–29.

Dalton, J., 1799: *Experiments and observations to determine whether the quantity of rain and dew is equal to the quantity of water carried off by the rivers and raised by evaporation: With an enquiry into the origin of springs*. R. & W. Dean, Manchster, 30 pp.

Darby, L. S., K. J. Allwine, and R. M. Banta, 2006: Nocturnal low-level jet in a mountain basin complex. Part II: Transport and diffusion of tracer under stable conditions. *J. Appl. Meteor. Climatol.*, **45 (5)**, 740–753.

Defant, F., 1950: Theorie der Land- und Seewinde. *Arch. Meteor. Geophys. Bioklim.*, **2 (4)**, 404–425.

Dick, G., G. Gendt, and C. Reigber, 2001: First experience with near real-time water vapor estimation in a German GPS network. *J. Atmos. Sol. Terr. Phys.*, **63 (12)**, 1295–1304.

Dingman, S., 2002: *Physical Hydrology*. 2d ed., Physical Hydrology, Prentice Hall, Upper Saddle River, New Jersey, 646 pp.

Dockery, D. W. and C. A. Pope, 1994: Acute respiratory effects of particulate air pollution. *Annu. Rev. Public Health*, **15 (1)**, 107–132.

Draper, N. R. and H. Smith, 1998: *Applied Regression Analys*. John Wiley & Sons, Hoboken, NJ, 312 pp.

Drexler, J. Z., R. L. Snyder, D. Spano, U. Paw, and K. Tha, 2004: A review of models and micrometeorological methods used to estimate wetland evapotranspiration. *Hydrol. Processes*, **18 (11)**, 2071–2101.

Duan, Z. and W. Bastiaanssen, 2015: A new empirical procedure for estimating intra-annual heat storage changes in lakes and reservoirs: review and analysis of 22 lakes. *Remote Sens. Environ.*, **156**, 143–156.

Durran, D. R., 1986: Another look at downslope windstorms. Part I: The development of analogs to supercritical flow in an infinitely deep, continuously stratified fluid. *J. Atmos. Sci.*, **43 (21)**, 2527–2543.

———, 2003: Downslope winds. *Encyclopedia of Atmospheric Sciences*, 644–650.

Dyer, A., 1974: A review of flux-profile relationships. *Boundary-Layer Meteorol.*, **7 (3)**, 363–372.

Emery, W., S. Castro, G. Wick, P. Schluessel, and C. Donlon, 2001: Estimating sea surface temperature from infrared satellite and in situ temperature data. *Bull. Am. Meteorol. Soc.*, **82 (12)**, 2773.

Etling, D., 2008: *Theoretische Meteorologie: Eine Einführung*. 3d ed., Springer, Berlin, Heidelberg, 376 pp.

Ficke, J. F., 1972: Comparison of evaporation computation methods, Pretty Lake, Lagrange County, northeastern Indiana. Tech. rep., U.S. Government Printing Office, Washington, D.C., 48 pp.

Fiedler, F., 1983: Einige Charakteristika der Strömung im Oberrheingraben. *Wissenschaftliche Berichte des Instituts für Meteorologie und Klimaforschung der Universität Karlsruhe*, **4**, 113–123.

Finnigan, J. J., R. Clement, Y. Malhi, R. Leuning, and H. Cleugh, 2003: A Re-Evaluation of Long-Term Flux Measurement Techniques Part I: Averaging and Coordinate Rotation. *Boundary-Layer Meteorol.*, **107 (1)**, 1–48.

Foken, T., 1999: Der Bayreuther Turbulenzknecht. Tech. Rep. 1, Univ. Bayreuth, Abt. Mikrometeorol., Bayreuth, 16 pp.

———, 2006: *Angewandte Meteorologie: Mikrometeorologische Methoden*. 2d ed., Springer, Berlin, 325 pp.

———, 2008: The energy balance closure problem: an overview. *Ecol. Appl.*, **18 (6)**, 1351–1367.

Foken, T., M. Aubinet, and R. Leuning, 2012a: The eddy covariance method. *Eddy covariance. A practical guide to measurements and data analysis*, Aubinet, M., T. Vesala, and D. Papale, Eds., Springer, Dordrecht, 1–19.

Foken, T., R. Leuning, S. Oncley, M. Mauder, and M. Aubinet, 2012b: Corrections and data quality control. *Eddy covariance. A practical guide to measurements and data analysis*, Aubinet, M., T. Vesala, and D. Papale, Eds., Springer, Dordrecht, 85–132.

Foken, T., G. Skeib, and S. Richter, 1991: Dependence of the integral turbulence characteristics on the stability of stratification and their use for Doppler-Sodar measurements. *Z. Meteorol.*, **41**, 311–115.

Foken, T. and B. Wichura, 1996: Tools for quality assessment of surface-based flux measurements. *Agric. For. Meteorol.*, **78 (1)**, 83 – 105.

Gavrieli, I., A. Bein, and A. Oren, 2005: The expected impact of the peace conduit project (the Red Sea-Dead Sea pipeline) on the Dead Sea. *Mitigation and Adaptation Strategies for Global Change*, **10 (1)**, 3–22.

Gendt, G., G. Dick, C. Reigber, M. Tomassini, Y. Liu, and M. Ramatschi, 2004: Near real time GPS water vapor monitoring for numerical weather prediction in Germany. *J. Meteor. Soc. Japan*, **82 (1B)**, 361–370.

Gertman, I. and A. Hecht, 2002: The Dead Sea hydrography from 1992 to 2000. *J. Mar. Sys.*, **35 (3–4)**, 169–181.

Givati, A. and A. Tal, 2016: Monthly hydrological status: Upper water in major drainage areas and water tables in the country system. The Hydrological Service and Water Authority, Israel, 27 pp.

Gohm, A. and G. Mayr, 2004: Hydraulic aspects of föhn winds in an Alpine valley. *Q. J. R. Meteorol. Soc.*, **130 (597)**, 449–480.

Goldreich, Y., 2003: *The climate of Israel : observation, research, and application*. Kluwer Academic/Plenum Publishers, New York [u.a.], 270 pp.

Goren, M. and R. Ortal, 1999: Biogeography, diversity and conservation of the inland water fish communities in Israel. *Biol. Conserv.*, **89 (1)**, 1 – 9.

Grubišic, V., J. D. Doyle, J. Kuettner, R. Dirks, S. A. Cohn, L. L. Pan, S. Mobbs, R. B. Smith, C. D. Whiteman, S. Czyzyk, et al., 2008: The Terrain-Induced Rotor Experiment: A field campaign overview including observational highlights. *Bull. Am. Meteorol. Soc.*, **89 (10)**, 1513–1533.

Gustavsson, T., M. Karlsson, J. Bogren, and S. Lindqvist, 1998: Development of temperature patterns during clear nights. *J. Appl. Meteorol.*, **37 (6)**, 559–571.

Haiden, T., 2003: On the pressure field in the slope wind layer. *J. Atmos. Sci.*, **60 (13)**, 1632–1635.

Hall, J. K., 1997: Topography and Bathymetry of the Dead Sea Depression. *The Dead Sea: the lake and its setting*, Niemi, T. M., Z. Ben-Avraham, and J. Gat, Eds., Oxford University Press, USA, Oxford, 11–21.

Harari, M., R. Barzillai, and J. Shani, 1998: Magnesium in the management of asthma: critical review of acute and chronic treatments, and Deutsches Medizinisches Zentrum's (DMZ's) clinical experience at the Dead Sea. *J. Asthma*, **35 (7)**, 525–536.

Harbeck, G., M. Kohler, and G. Koberg, 1952: Water loss investigations: Lake Hefner studies. *US Geol. Surv. Prof Pap*, **269**, 158.

Harbeck, G. E., 1962: A practical field technique for measuring reservoir evaporation utilizing mass-transfer theory. *Geological Survy Professional Paper*, **272-E**, 101 – 105.

Hecht, A. and I. Gertman, 2003: Dead Sea meteorological climate. *Fungal Life in the Dead Sea*, Nevo, E., A. Oren, and S. Wasser, Eds., International Center for Cryptogamic Plants and Fungi, Haifa, 68–114.

Henderson-Sellers, B., 1986: Calculating the surface energy balance for lake and reservoir modeling: A review. *Rev. Geophys.*, **24 (3)**, 625–649.

Heusinkveld, B., A. Jacobs, A. Holtslag, and S. Berkowicz, 2004: Surface energy balance closure in an arid region: role of soil heat flux. *Agric. For. Meteorol.*, **122 (1)**, 21–37.

Højstrup, J., 1981: A simple model for the adjustment of velocity spectra in unstable conditions downstream of an abrupt change in roughness and heat flux. *Boundary-Layer Meteorol.*, **21 (3)**, 341–356.

Holtzman, R., U. Shavit, M. Segal-Rozenhaimer, I. Gavrieli, A. Marei, E. Farber, and A. Vengosh, 2005: Quantifying ground water inputs along the lower Jordan River. *J. Environ. Qual.*, **34 (3)**, 897–906.

Horst, T. and J. Doran, 1988: The turbulence structure of nocturnal slope flow. *J. Atmos. Sci.*, **45 (4)**, 605–616.

Ilotoviz, E., H. Shafir, P. Gasch, and P. Alpert, 2015: Semi operational prediction of the Dead Sea evaporation - A synoptic systems approach. *J. Water Resour. Prot.*, **7 (13)**, 1058.

Inagaki, A., M. O. Letzel, S. Raasch, and M. Kanda, 2006: Impact of surface heterogeneity on energy imbalance: a study using LES. *J. Meteor. Soc. Japan*, **84 (1)**, 187–198.

Jackson, P. L., G. Mayr, and S. Vosper, 2013: Dynamically-driven winds. *Mountain Weather Research and Forecasting: Recent Progress and Current Challenges*, Chow, K. F., F. S. De Wekker, and J. B. Snyder, Eds., Springer Netherlands, Dordrecht, 121–218.

Jacobs, A. F., B. G. Heusinkveld, and S. M. Berkowicz, 1999: Dew deposition and drying in a desert system: a simple simulation model. *J. Arid. Environ.*, **42 (3)**, 211–222.

———, 2002: A simple model for potential dewfall in an arid region. *Atmos. Res.*, **64 (1)**, 285–295.

Jaffe, S., 1977: Number of seasons in the year: a climatic point of view. *Meteorologia BeIsrael*, **33 (2)**, 11 – 15.

Jiang, Q. and J. D. Doyle, 2008: Diurnal variation of downslope winds in Owens Valley during the Sierra Rotor Experiment. *Mon. Weather Rev.*, **136 (10)**, 3760–3780.

Jonsson, A., J. Åberg, A. Lindroth, and M. Jansson, 2008: Gas transfer rate and CO2 flux between an unproductive lake and the atmosphere in northern Sweden. *J. Geophys. Res. Biogeosci.*, **113 (G4)**.

Kaimal, J., J. Wyngaard, Y. Izumi, and O. Coté, 1972: Spectral characteristics of surface-layer turbulence. *Q. J. R. Meteorol. Soc.*, **98 (417)**, 563–589.

Kalthoff, N., B. Adler, A. Wieser, M. Kohler, K. Träumner, J. Handwerker, U. Corsmeier, S. Khodayar, D. Lambert, A. Kopmann, et al., 2013: KITcube–a mobile observation platform for convection studies deployed during HyMeX. *Meteorol. Z.*, **22 (6)**, 633–647.

Kalthoff, N., M. Fiebig-Wittmaack, C. Meißner, M. Kohler, M. Uriarte, I. Bischoff-Gauss, and E. Gonzales, 2006: The energy balance, evapotranspiration and nocturnal dew deposition of an arid valley in the Andes. *J. Arid. Environ.*, **65 (3)**, 420–443.

Kalthoff, N., V. Horlacher, U. Corsmeier, A. Volz-Thomas, B. Kolahgar, H. Geiß, M. Möllmann-Coers, and A. Knaps, 2000: Influence of valley winds on transport and dispersion of airborne pollutants in the Freiburg-Schauinsland area. *J. Geophys. Res. Atmos*, **105 (D1)**, 1585–1597.

Kalthoff, N. and B. Vogel, 1992: Counter-current and channelling effect under stable stratification in the area of Karlsruhe. *J. Arid. Environ.*, **45**, 113–126.

Kampf, S. K. and S. W. Tyler, 2006: Spatial characterization of land surface energy fluxes and uncertainty estimation at the Salar de Atacama, Northern Chile. *Adv. Water Res.*, **29 (2)**, 336–354.

Kanda, M., A. Inagaki, M. O. Letzel, S. Raasch, and T. Watanabe, 2004: LES study of the energy imbalance problem with eddy covariance fluxes. *Boundary-Layer Meteorol.*, **110 (3)**, 381–404.

Kanemasu, E., S. Verma, E. Smith, L. Fritschen, M. Wesely, R. Field, W. Kustas, H. Weaver, J. Stewart, R. Gurney, et al., 1992: Surface flux measurements in FIFE: An overview. *J. Geophys. Res. Atmos*, **97 (D17)**, 18 547–18 555.

Khlaifat, A., M. Hogan, G. Phillips, K. Nawayseh, J. Amira, and E. Talafeha, 2010: Long-term monitoring of the Dead Sea level and brine physico-chemical parameters from 1987 to 2008. *J. Mar. Sys.*, **81 (3)**, 207 – 212.

Kidron, G. J., 2000: Analysis of dew precipitation in three habitats within a small arid drainage basin, Negev Highlands, Israel. *Atmos. Res.*, **55 (3–4)**, 257–270.

Kidron, G. J., I. Herrnstadt, and E. Barzilay, 2002: The role of dew as a moisture source for sand microbiotic crusts in the Negev Desert, Israel. *J. Arid. Environ.*, **52 (4)**, 517–533.

Kishcha, P., D. Rieger, J. Metzger, B. Starobinets, M. Bangert, H. Vogel, U. Schï¿½ttler, U. Corsmeier, P. Alpert, and B. Vogel, 2016: Modelling of a strong dust event in the complex terrain of the Dead Sea valley during the passage of a gust front. *Tellus B*, **68**, 1 – 19.

Klemp, J. and D. Durran, 1987: Numerical modelling of bora winds. *Meteorol. Atmos. Phys.*, **36 (1-4)**, 215–227.

Klemp, J. and D. Lilly, 1975: The dynamics of wave-induced downslope winds. *J. Atmos. Sci.*, **32 (2)**, 320–339.

Kljun, N., P. Calanca, M. W. Rotach, and H. P. Schmid, 2015: A simple two-dimensional parameterisation for Flux Footprint Prediction (FFP). *Geosci. Model Dev.*, **8 (11)**, 3695–3713.

Kohler, M. and L. Parmele, 1967: Generalized estimates of free-water evaporation. *Water Resour. Res.*, **3 (4)**, 997–1005.

Konda, M., N. Imasato, K. Nishi, and T. Toda, 1994: Measurement of the sea surface emissivity. *J. Oceanogr.*, **50 (1)**, 17–30.

Kondrat'Ev, K. Y., 1969: *Radiation in the Atmosphere*. International Geophysics Series, Vol. 12, Academic Press, New York, 915 pp.

Kottmeier, C., A. Agnon, D. Al-Halbouni, P. Alpert, U. Corsmeier, T. Dahm, A. Eshel, S. Geyer, M. Haas, E. Holohan, et al., 2016: New perspectives on interdisciplinary earth science at the Dead Sea: The DESERVE project. *Sci. Total Environ.*, **544**, 1045–1058.

Kottmeier, C., P. Palacio-Sese, N. Kalthoff, U. Corsmeier, and F. Fiedler, 2000: Sea breezes and coastal jets in southeastern Spain. *Int. J. Climatol.*, **20 (14)**, 1791–1808.

Krishnan, P., T. P. Meyers, R. L. Scott, L. Kennedy, and M. Heuer, 2012: Energy exchange and evapotranspiration over two temperate semi-arid grasslands in North America. *Agric. For. Meteorol.*, **153**, 31–44.

Kühnlein, C., A. Dörnbrack, and M. Weissmann, 2013: High-resolution Doppler lidar observations of transient downslope flows and rotors. *Mon. Weather Rev.*, **141 (10)**, 3257–3272.

Lehner, M., C. D. Whiteman, S. W. Hoch, E. T. Crosman, M. E. Jeglum, N. W. Cherukuru, R. Calhoun, B. Adler, N. Kalthoff, R. Rotunno, et al., 2016: The METCRAX II field experiment: A study of downslope windstorm-type flows in Arizonas Meteor Crater. *Bulletin of the American Meteorological Society*, **97 (2)**, 217–235.

Lensky, N. G., Y. Dvorkin, V. Lyakhovsky, I. Gertman, and I. Gavrieli, 2005: Water, salt, and energy balances of the Dead Sea. *Water Resour. Res.*, **41 (12)**, 1–13.

Leuning, R., O. Denmead, A. Lang, and E. Ohtaki, 1982: Effects of heat and water vapor transport on eddy covariance measurement of CO2 fluxes. *Boundary-Layer Meteorol.*, **23 (2)**, 209–222.

Lipchin, C., D. Sandler, and E. Cushman, 2009: *The Jordan River and Dead Sea Basin: Cooperation Amid Conflict*. Springer Science & Business Media, 315 pp.

Long, R. R., 1954: Some Aspects of the Flow of Stratified Fluids: II. Experiments with a Two-Fluid System. *Tellus*, **6 (2)**, 97–115.

Mahrer, Y. and R. Pielke, 1977: The effects of topography on sea and land breezes in a two-dimensional numerical model. *Mon. Weather Rev.*, **105**, 1151 – 1162.

Mahrt, L., 1998: Flux sampling errors for aircraft and towers. *J. Atmos. Oceanic Technol.*, **15 (2)**, 416–429.

———, 2006: Variation of surface air temperature in complex terrain. *J. Appl. Meteor. Climatol.*, **45 (11)**, 1481–1493.

Malek, E., G. McCurdy, and B. Giles, 1999: Dew contribution to the annual water balances in semi-arid desert valleys. *J. Arid. Environ.*, **42 (2)**, 71–80.

Mammarella, I., A. Nordbo, Ü. Rannik, S. Haapanala, J. Levula, H. Laakso, A. Ojala, O. Peltola, J. Heiskanen, J. Pumpanen, et al., 2015: Carbon dioxide and energy fluxes over a small boreal lake in Southern Finland. *J. Geophys. Res. Biogeosci.*, **120 (7)**, 1296–1314.

Mauder, M., M. Cuntz, C. Drüe, A. Graf, C. Rebmann, H. P. Schmid, M. Schmidt, and R. Steinbrecher, 2013: A strategy for quality and uncertainty assessment of long-term eddy-covariance measurements. *Agric. For. Meteorol.*, **169**, 122 – 135.

Mauder, M. and T. Foken, 2011: *Documentation and instruction manual of the eddy-covariance software package TK3*, Vol. 46. Univ. Bayreuth, Abt. Mikrometeorologie, 60 pp.

Mayr, G. J. and L. Armi, 2008: Föhn as a response to changing upstream and downstream air masses. *Q. J. R. Meteorol. Soc.*, **134 (635)**, 1357–1369.

———, 2010: The influence of downstream diurnal heating on the descent of flow across the Sierras. *J. Appl. Meteor. Climatol.*, **49 (9)**, 1906–1912.

McKee, T. B. and R. D. O'Neal, 1989: The role of valley geometry and energy budget in the formation of nocturnal valley winds. *J. Appl. Meteorol.*, **28 (6)**, 445–456.

Miller, S. T. K., 2003: Sea breeze: Structure, forecasting, and impacts. *Rev. Geophys.*, **41 (3)**, 1–31.

Montgomery, R., 1948: Vertical eddy flux of heat in the atmosphere. *J. Meteor.*, **5 (6)**, 265–274.

Moore, C., 1986: Frequency response corrections for eddy correlation systems. *Boundary-Layer Meteorol.*, **37 (1-2)**, 17–35.

Morton, F. I., 1983: Operational estimates of lake evaporation. *J. Hydrol.*, **66 (1-4)**, 77–100.

Moses, S. W., M. David, E. Goldhammer, A. Tal, and S. Sukenik, 2006: The Dead Sea, a unique natural health resort. *IMAJ–RAMAT GAN*, **8 (7)**, 483.

Naor, R., O. Potchter, H. Shafir, and P. Alpert, 2015: An observational study of the summer Mediterranean Sea breeze front penetration into the complex topography of the Jordan Rift Valley. *Theor. Appl. Climatol.*, 1–10.

Nehorai, R., N. Lensky, S. Brenner, and I. Lensky, 2013: The dynamics of the skin temperature of the Dead Sea. *Advances in Meteorology*, **2013**, 1–9.

Nordbo, A., S. Launiainen, I. Mammarella, M. Leppäranta, J. Huotari, A. Ojala, and T. Vesala, 2011: Long-term energy flux measurements and energy balance over a small boreal lake using eddy covariance technique. *J. Geophys. Res. Atmos*, **116 (D2)**.

Obukhov, A., 1951: Investigation of the micro-structure of the wind in the near-surface layer of the atmosphere. *Izvestia AN SSSR, seria Geofizika*, **3**, 49ff.

Oroud, I., 1995: Effects of salinity upon evaporation from pans and shallow lakes near the Dead Sea. *Theor. Appl. Climatol.*, **52 (3-4)**, 231–240.

———, 2011: Evaporation estimates from the Dead Sea and their implications on its water balance. *Theor. Appl. Climatol.*, **106 (3-4)**, 523–530.

Ostro, B., 1993: The association of air pollution and mortality: examining the case for inference. *Archives of Environmental Health: An International Journal*, **48 (5)**, 336–342.

Ostro, B. D., S. Hurley, and M. J. Lipsett, 1999: Air oollution and daily mortality in the Coachella Valley, California: A study of PM10 dominated by coarse particles. *Environ. Res.*, **81 (3)**, 231 – 238.

Peltier, W. and T. Clark, 1979: The evolution and stability of finite-amplitude mountain waves. Part II: Surface wave drag and severe downslope windstorms. *J. Atmos. Sci.*, **36 (8)**, 1498–1529.

Penman, H. L., 1948: Natural evaporation from open water, bare soil and grass. *Proceedings of the Royal Society of London A: Mathematical, Physical and Engineering Sciences*, The Royal Society, Vol. 193, 120–145.

Picard, R. R. and R. D. Cook, 1984: Cross-validation of regression models. *J. Am. Stat. Assoc.*, **79 (387)**, 575–583.

Plavcan, D., G. J. Mayr, and A. Zeileis, 2014: Automatic and probabilistic foehn diagnosis with a statistical mixture model. *J. Appl. Meteor. Climatol.*, **53 (3)**, 652–659.

Porson, A., D. Steyn, and G. Schayes, 2007a: Sea-breeze scaling from numerical model simulations, part I: Pure sea breezes. *Boundary-Layer Meteorol.*, 17–29.

Porson, A., D. G. Steyn, and G. Schayes, 2007b: Sea-breeze scaling from numerical model simulations, part II: Interaction between the sea breeze and slope flows. *Boundary-Layer Meteorol.*, 31–41.

Prandtl, L., K. Oswatitsch, and K. Wieghardt, 1942: *Fuehrer durch die Strömungslehre*. Vieweg and Sohn, Braunschweig, 382 pp.

Priestley, C. H. B., 1972: On the assessment of surface heat flux and evaporation using large scale parameters. *Mon. Weather Rev*, **100 (2)**, 81–92.

Ramis, C. and R. Romero, 1995: A first numerical simulation of the development and structure of the sea breeze on the Island of Mallorca. *Ann. Geophys.*, **13 (9)**, 981–994.

Rampanelli, G. and D. Zardi, 2004: A method to determine the capping inversion of the convective boundary layer. *J. Appl. Meteorol.*, **43 (6)**, 925–933.

Rasmussen, A. H., M. Hondzo, and H. G. Stefan, 1995: A Test of Several Evaporation Equations for Water Temperature Simulations in Lakes. *J. Am. Water Resour. Assoc.*, **31 (6)**, 1023–1028.

Rebmann, C., O. Kolle, B. Heinesch, R. Queck, A. Ibrom, and M. Aubinet, 2012: Data Acquisition and Flux Calculations. *Eddy covariance. A practical guide to measurements and data analysis*, Aubinet, M., T. Vesala, and D. Papale, Eds., Springer, Dordrecht, 59–84.

Rocken, C., T. V. Hove, J. Johnson, F. Solheim, R. Ware, M. Bevis, S. Chiswell, and S. Businger, 1995: GPS/STORM–GPS sensing of atmospheric water vapor for meteorology. *J. Atmos. Oceanic Technol.*, **12 (3)**, 468–478.

Rocken, C., T. Van Hove, and R. Ware, 1997: Near real-time GPS sensing of atmospheric water vapor. *Geophys. Res. Lett.*, **24 (24)**, 3221–3224.

Rosenberry, D. O., T. C. Winter, D. C. Buso, and G. E. Likens, 2007: Comparison of 15 evaporation methods applied to a small mountain lake in the northeastern USA. *J. Hydrol.*, **340 (3)**, 149–166.

Sakai, R. K., D. R. Fitzjarrald, and K. E. Moore, 2001: Importance of low-frequency contributions to eddy fluxes observed over rough surfaces. *J. Appl. Meteorol.*, **40 (12)**, 2178–2192.

Salameh, E., 1996: Water quality degradation in Jordan (Impacts on Environment, Economy and Future Generations Resources Base). *Friedrich Ebert Stiftung, Royal Society for the Conservation of Nature.*

Salameh, E. and H. El-Naser, 1999: Does the actual drop in Dead Sea level reflect the development of water sources within its drainage basin? *Acta Hydrochim. Hydrobiol.*, **27 (1)**, 5–11.

———, 2000: Changes in the Dead Sea Level and their Impacts on the Surrounding Groundwater Bodies. *Acta Hydrochim. Hydrobiol.*, **28 (1)**, 24–33.

Salhotra, A. M., E. E. Adams, and D. R. F. Harleman, 1985: Effect of Salinity and Ionic Composition on Evaporation: Analysis of Dead Sea Evaporation Pans. *Water Resour. Res.*, **21 (9)**, 1336–1344.

Savijärvi, H. and M. Alestalo, 1988: The sea breeze over a lake or gulf as the function of the prevailing flow. *Beitr. Phys. Atmos*, **61 (2)**, 98–104.

Schmidli, J. and R. Rotunno, 2010: Mechanisms of along-valley winds and heat exchange over mountainous terrain. *J. Atmos. Sci.*, **67 (9)**, 3033–3047.

Schotanus, P., F. Nieuwstadt, and H. De Bruin, 1983: Temperature measurement with a sonic anemometer and its application to heat and moisture fluxes. *Boundary-Layer Meteorol.*, **26 (1)**, 81–93.

Shafir, H. and P. Alpert, 2011: Regional and local climatic effects on the Dead-Sea evaporation. *Climatic Change*, **105 (3-4)**, 455–468.

Shafir, H., F. J. Jin, C. M. Lati, Y., and P. Alpert, 2008: Wind Channeling by the Dead-Sea Wadis. *The Open Atmospheric Science Journal*, **2**, 139–152.

Siebert, C., T. Rödiger, S. Geyer, J. B. Laronne, N. Hillel, M. Sauter, and U. Mallast, 2016: Multidisciplinary investigations of the transboundary Dead Sea basin and its water resources. *Integrated Water Resources Management: Concept, Research and Implementation*, Springer, 107–127.

Siebert, C., T. Rödiger, U. Mallast, A. Gräbe, J. Guttman, J. B. Laronne, Y. Storz-Peretz, A. Greenman, E. Salameh, M. Al-Raggad, D. Vachtman, A. B. Zvi, D. Ionescu, A. Brenner, R. Merz, and S. Geyer, 2014: Challenges to estimate surface- and groundwater flow in arid regions: The Dead Sea catchment. *Sci. Total Environ.*, **485?486 (0)**, 828 – 841.

Simpson, J. E., 1994: *Sea breeze and local winds*. Cambridge Univ. Pr, Cambridge [u.a.], 240 pp.

Smiatek, G., H. Kunstmann, and A. Heckl, 2011: High-resolution climate change simulations for the Jordan River area. *J. Geophys. Res. Atmos*, **116 (D16)**.

Stanhill, G., 1987: The radiation climate of the dead sea. *J. Climatol.*, **7 (3)**, 247–265.

———, 1994: Changes in the rate of evaporation from the Dead Sea. *Int. J. Climatol.*, **14 (4)**, 465–471.

Steinacker, R., 1984: Area-height distribution of a valley and its relation to the valley wind. *Beiträge zur Physik der Atmosphäre*, **57 (1)**, 64–71.

Steinhorn, I., G. Assaf, J. Gat, A. Nishry, A. Nissenbaum, M. Stiller, M. t. Beyth, D. Neev, R. Garber, G. Friedman, et al., 1979: The Dead Sea: deepening of the mixolimnion signifies the overture to overturn of the water column. *Science*, **206 (4414)**, 55–57.

Stone, M., 1974: Cross-validatory choice and assessment of statistical predictions. *Journal of the Royal Statistical Society. Series B (Methodological)*, 111–147.

Stull, R. B., 1988: *An introduction to boundary layer meteorology*. Repr. 2009 ed., Springer, Dordrecht, 670 pp.

Sturman, A. P., H. A. McGowan, and R. A. Spronken-Smith, 1999: Mesoscale and local climates in New Zealand. *Prog. Phys. Geog.*, **23 (4)**, 611–635.

Sugita, M., J. Asanuma, M. Tsujimura, S. Mariko, M. Lu, F. Kimura, D. Azzaya, and T. Adyasuren, 2007: An overview of the rangelands atmosphere–hydrosphere–biosphere interaction study experiment in northeastern Asia (RAISE). *J. Hydrol.*, **333 (1)**, 3–20.

Sukenik, S., 1996: Balneotherapy for rheumatic diseases at the Dead Sea area. *Isr. J. Med. Sci.*, **32**, S16–9.

Swinbank, W., 1951: The measurement of vertical transfer of heat and water vapor by eddies in the lower atmosphere. *J. Meteor.*, **8 (3)**, 135–145.

Tal, A., 2006: Seeking sustainability: Israel's evolving water management strategy. *Science*, **313 (5790)**, 1081–1084.

Thomas, C. and T. Foken, 2002: Re-evaluation of integral turbulence characteristics and their parameterisations. *15th Conference on Boundary Layer and Turbulence*, 15–19.

Träumner, K., 2012: *Einmischprozesse am Oberrand der konvektiven atmosphärischen Grenzschicht*. Wissenschaftliche Berichte des Instituts für Meteorologie und Klimaforschung des Karlsruher Instituts für Technologie ; Vol. 51, KIT Scientific Publishing, Karlsruhe, zugl.: Karlsruhe, KIT, Diss., 2010.

Träumner, K., A. Wieser, B. Ruck, C. Frank, L. Röhner, and C. Kottmeier, 2012: The suitability of Doppler lidar for characterizing the wind field above forest edges. *Forestry*, **85 (3)**, 399–412.

Van Bavel, C., 1966: Potential evaporation: the combination concept and its experimental verification. *Water Resour. Res.*, **2 (3)**, 455–467.

Vergeiner, I. and E. Dreiseitl, 1987: Valley winds and slope winds - observations and elementary thoughts. *Meteorol. Atmos. Phys.*, **36 (1-4)**, 264–286.

Vogel, B., C. Hoose, H. Vogel, and C. Kottmeier, 2006: A model of dust transport applied to the Dead Sea Area. *Meteorol. Z.*, **15 (6)**, 611–624.

Wagner, A., 1932: Neue Theorie des Berg-und Talwindes. *Meteorol. Z.*, **49 (9)**.

Webb, E. K., G. I. Pearman, and R. Leuning, 1980: Correction of flux measurements for density effects due to heat and water vapour transfer. *Q. J. R. Meteorol. Soc.*, **106 (447)**, 85–100.

Weigel, A. P., F. K. Chow, and M. W. Rotach, 2007: On the nature of turbulent kinetic energy in a steep and narrow Alpine valley. *Boundary-Layer Meteorol.*, **123 (1)**, 177–199.

Wexler, R., 1946: Theory and observations of land and sea breezes. *Bull. Amer. Meteor. Soc*, **27**, 272–287.

Whiteman, C., 2000: *Mountain Meteorology: fundamentals and applications*. Oxford Univ. Press, New York, 376 pp.

Whiteman, C. D. and J. C. Doran, 1993: The relationship between overlying synoptic-scale flows and winds within a valley. *J. Appl. Meteorol.*, **32 (11)**, 1669–1682.

Wilczak, J., S. Oncley, and S. Stage, 2001: Sonic anemometer tilt correction algorithms. *Boundary-Layer Meteorol.*, **99 (1)**, 127–150.

Wilson, K., A. Goldstein, E. Falge, M. Aubinet, D. Baldocchi, P. Berbigier, C. Bernhofer, R. Ceulemans, H. Dolman, C. Field, et al., 2002: Energy balance closure at FLUXNET sites. *Agric. For. Meteorol.*, **113 (1)**, 223–243.

Winter, T., D. Rosenberry, and A. Sturrock, 1995: Evaluation of 11 equations for determining evaporation for a small lake in the north central United States. *Water Resour. Res.*, **31 (4)**, 983–993.

Wippermann, F., 1984: Air flow over and in broad valleys: Channeling and counter-current. *Beiträge zur Physik der Atmosphäre*, **57 (1)**, 92–105.

Xu, Q.-S., Y.-Z. Liang, and Y.-P. Du, 2004: Monte Carlo cross-validation for selecting a model and estimating the prediction error in multivariate calibration. *J. Chemom.*, **18 (2)**, 112–120.

Yechieli, Y., M. Abelson, and G. Baer, 2016: Sinkhole formation and subsidence along the Dead Sea coast, Israel. *Hydrol. J.*, **24 (3)**, 601–612.

Yechieli, Y., M. Abelson, A. Bein, O. Crouvi, and V. Shtivelman, 2006: Sinkhole swarms along the Dead Sea coast: reflection of disturbance of lake and adjacent groundwater systems. *Geol. Soc. Am. Bull.*, **118 (9-10)**, 1075–1087.

Zardi, D. and C. D. Whiteman, 2013: Diurnal Mountain Wind Systems. *Mountain Weather Research and Forecasting: Recent Progress and Current Challenges*, Chow, K. F., F. S. De Wekker, and J. B. Snyder, Eds., Springer Netherlands, Dordrecht, 35–119.

Zhong, S., J. Li, C. D. Whiteman, X. Bian, and W. Yao, 2008: Climatology of high wind events in the Owens Valley, California. *Mon. Weather Rev.*, **136 (9)**, 3536–3552.

Glossary

α	albedo	21
β	Bowen ratio	26
γ	psychometric constant	27
ε	emissivity coefficient	22
$\eta(x)$	height of free surface in x direction	17
$\overline{\Theta}(z)$	mean potential temperature profile	16
$\overline{\Theta}$	mean potential temperature	17
κ	Kármán constant	25
μ	ratio of molar masses of dry air and water vapour	56
$\Pi(T_m)$	conversion factor for IWV determination	42
ρ_a	air density	24
ρ_w	water density	25
σ_N	standard deviation of the noise	45
σ	ratio of water vapour and dry air densities	56
σ_s	standard deviation of the received signal strength	45
φ_H	flux-profile functions after Dyer	65
Ψ_H	integral over the flux-profile functions	65
a_G	molecular heat transfer coefficient of the soil	23
a_{sw}	activity of the saline water	64
c	phase speed	45
$S_{x(y)}$	theoretical (ideal) form of the spectrum	58
c_p	specific heat capacity at constant pressure	23
c_{PT}	Priestley-Taylor coefficient	29

D_H	turbulent transfer coefficient for sensible heat flux	23
D_{LE}	turbulent transfer coefficient for latent heat flux	23
e_a	vapour pressure of air	25
Δe	vapour pressure deficit	63
Δe_{T_w}	vapour pressure deficit calculated with T_w	66
Δe_{MO}	vapour pressure deficit calculated with T_{MO}	66
er	prediction error	69
E_{sw}	saturation vapour pressure of the saline water	64
Ev	evaporation	24
E_w	saturation vapour pressure of water	25
f	frequency	45
f_c	Coriolis parameter	10
F_n	net heat flux associated with water in- and outflow and precipitation	22
Fr	Froude number	18
F_v	water vapour flux	55
g	gravity acceleration	16
G	ground heat flux	22
H	sensible heat flux	22
$\hat{H}$	non-dimensional mountain height	17
ΔH_v	molar latent heat of vaporisation	62
h_c	crest height	17
$h(x)$	terrain height in x direction	17
k_B	Stephan-Boltzmann constant	22
K_E	bulk transfer coefficient for evaporation	25
$K\downarrow$	solar radiation flux	21
$L\downarrow$	longwave downward radiation	22
LE	latent heat flux	22
L	net longwave radiation	29
L_*	Monin-Obukhov Length	65
$L\uparrow$	longwave upward radiation	22

L_v	latent heat of vaporisation	23
N	Brunt-Vaisala frequency	17
p	air pressure	25
q	specific humidity	23
ΔQ	storage term	22
Ri_b	Bulk-Richardson number	26
Rn	net radiation	22
R^2	coefficient of determination	68
T_a	air temperature	23
T_B	thermodynamic temperature	22
$T_{x(y)}$	specific transfer function	58
T_m	meant temperature of the atmosphere	42
T_{soil}	soil temperature	23
T_s	surface temperature	22
T_w	water temperature	23
T_{MO}	surface water temperature obtained by Monin-Obukhov theory	64
$\Delta trop$	tropospheric delay	42
u_*	friction velocity	59
u_{up}	upstream wind velocity	17
$\overline{u}(z)$	mean profile of horizontal wind component u	16
v_a	wind speed	25
v_r	radial velocity	45
$\overline{v}(z)$	mean profile of horizontal wind component v	16
z	height above ground	17
z_0	roughness length	25
z_d	zero-plane displacement height	25
z_m	measurement height	25

Acronyms

ABL	Atmospheric Boundary Layer	9
AGL	Above Ground Level	12
AMSL	Above Mean Sea Level	1
DESERVE	DEad SEa Rsearch VEnue	2
DWF	downslope windstorm-type flow	ii
EBC	Eenergy Balance Closure	134
EBS	Energy Balance Station	38
GFS	Global Forecast System	114
GFZ	German Research Centre of Geoscience	41
GPS	Global Positioning System	41
IMS	Israel Meteorological Service	38
IOP	Intensive Observation Period	46
IRGASON	Integrated Gas Analyser and Sonic Anemometer	40
ITC	Integral Turbulence Characteristics	58
IWV	integrated water vapour	41
LT	Local Time (UTC + 2 h)	46
MAD	Median Absolute Deviation	53
MAP	Mesoscale Alpine Programme	4
MCCV	Monte-Carlo cross validation	69
MET	Meteorological Tower	38
MSB	Mediterranean Sea Breeze	2
PPI	Plan-Position-Indicator	45
RHI	Range-Height-Indicator	45

RST	Red Sea Trough	36
SNR	Signal-to-Noise Ratio	44
SOP	Special Observation Period	8
TAF	Topographic Amplification Factor	13
TKE	Turbulent Kinetic Energy	12
TREX	Terrain-induced Rotor EXperiment	4
UTC	Coordinated Universal Time	46
VAD	Velocity-Azimuth-Display	45
ZHD	Zenith Hydrostatic Delay	42
ZTD	Zenith Total Delay	42
ZWD	Zenith Wet Delay	42

Acknowledgement

I would like to thank my supervisor Prof. Dr. Kottmeier to have made this Ph.D thesis possible and for his continuous support and advice. Furthermore, I want to thank PD Dr. Michael Kunz for co-examining this thesis.

I am particularly thankful to Dr. Ulrich Corsmeier for the opportunity to individually shape my work, for his excellent supervision, advice, and help throughout my Ph.D., and the opportunity to participate in several field campaigns.

I'm grateful to my colleagues and friends, Manuela Nied and Friederike Lott, for the fruitful discussions, critical reviews, coffee breaks, and sharing tea. Thank you both for making business trips more pleasant and of course for the joint spare time activities.

As it is the nature of an experimental doctoral thesis this work would not have been possible without the help of many colleagues. I want to thank everybody involved in the HEADS 2014 campaign. In particular, I want to thank Bernhard Deny and Philipp Gasch for their continuous support and help with the measurements and maintenance of the stations. Furthermore, I'm grateful to Martin Kohler, Norbert Kalthoff, Andreas Wieser, Jan Handwerker, and the technical staff at IMK for their great efforts to deploy the KITcube under the exceptional harsh climatic and difficult political conditions at the Dead Sea. In addition I wish to thank our Israeli cooperation partners from Tel Aviv University, Prof. Dr. Pinhas Alpert and his working group for their assistance with the realisation of the measurement campaign, Eduard Karat for managing the import/export formalities, and Yosi Lathi for his advice and help in so many different things. Moreover, thanks to Ethan Campbell from the National Parks

Authority Masada for providing the measurement site, Galina Dick from GFZ Potsdam for post-processing the GPS data, Holger Mahlke and Rainer Behrendt from Wetter3.de for providing GFS model data, and Heike Vogel for making the COSMO-EU analysis data available. Special thanks to Prof. Dr. Jörg Kleffmann for the opportunity to perform the laboratory measurements, his help, and for sharing his expertise with me.

I also wish to thank Bianca Adler for her advice concerning the data analysis and the helpful discussions, Gabi Klinck for her technical support, Rosalba Gräbner for her help with the administrative things, Beata Czajka for proof-reading the thesis and of course I want to thank the rest of my colleagues at the institute for creating such a pleasant working environment. I really enjoy working at IMK.

The last years were made enjoyable in large part due to my friends. Thank you for the joint vacations, short trips, nights out, dinners, phone calls, and the occasional after-work drink. Thank you for listening, offering me advice, and supporting me through this entire process.

A special thanks goes to Felix Vüllers. Dear Felix, I can not express how grateful I am for everything you've done for me. I'm especially thankful for your support, your motivating and encouraging words in the last months and for always cheering me up when I bothered too much about the thesis.

Finally, I want to thank my family. Especially my parents, Irmgard and Rolf Metzger, for their unconditional love and support. You always backed my decisions and helped me with your advice. I'm grateful for everything you did and made possible for me. Also a big thank you to my brothers, Axel Metzger and Felix Metzger, my sister Ute Pape, and their families. I always enjoy spending time with you, love seeing my nieces and nephew grow up, and I'm grateful about the much needed distraction from work. Especially your visits in Karlsruhe were always a highlight.

Wissenschaftliche Berichte des Instituts für Meteorologie und Klimaforschung des Karlsruher Instituts für Technologie (0179-5619)

Bisher erschienen:

Nr. 1: *Fiedler, F. / Prenosil, T.*
Das MESOKLIP-Experiment. (Mesoskaliges Klimaprogramm im Oberrheintal). August 1980

Nr. 2: *Tangermann-Dlugi, G.*
Numerische Simulationen atmosphärischer Grenzschichtströmungen über langgestreckten mesoskaligen Hügelketten bei neutraler thermischer Schichtung. August 1982

Nr. 3: *Witte, N.*
Ein numerisches Modell des Wärmehaushalts fließender Gewässer unter Berücksichtigung thermischer Eingriffe. Dezember 1982

Nr. 4: *Fiedler, F. / Höschele, K. (Hrsg.)*
Prof. Dr. Max Diem zum 70. Geburtstag. Februar 1983 (vergriffen)

Nr. 5: *Adrian, G.*
Ein Initialisierungsverfahren für numerische mesoskalige Strömungs modelle. Juli 1985

Nr. 6: *Dorwarth, G.*
Numerische Berechnung des Druckiderstandes typischer Geländeformen. Januar 1986

Nr. 7: *Vogel, B.; Adrian, G. / Fiedler, F.*
MESOKLIP-Analysen der meteorologischen Beobachtungen von mesoskaligen Phänomenen im Oberrheingraben. November 1987

Nr. 8: *Hugelmann, C.-P.*
Differenzenverfahren zur Behandlung der Advektion. Februar 1988

Nr. 9: *Hafner, T.*
Experimentelle Untersuchung zum Druckwiderstand der Alpen. April 1988

Nr. 10: *Corsmeier, U.*
Analyse turbulenter Bewegungsvorgänge in der maritimen atmosphärischen Grenzschicht. Mai 1988

Nr. 11: *Walk, O. / Wieringa, J.(eds)*
Tsumeb Studies of the Tropical Boundary-Layer Climate. Juli 1988

Nr. 12: *Degrazia, G. A.*
Anwendung von Ähnlichkeitsverfahren auf die turbulente Diffusion in der konvektiven und stabilen Grenzschicht. Januar 1989

Nr. 13: *Schädler, G.*
Numerische Simulationen zur Wechselwirkung zwischen Landoberflächen und atmophärischer Grenzschicht. November 1990

Nr. 14: *Heldt, K.*
Untersuchungen zur Überströmung eines mikroskaligen Hindernisses in der Atmosphäre. Juli 1991

Nr. 15: *Vogel, H.*
Verteilungen reaktiver Luftbeimengungen im Lee einer Stadt – Numerische Untersuchungen der relevanten Prozesse. Juli 1991

Nr. 16: *Höschele, K.(ed.)*
Planning Applications of Urban and Building Climatology – Proceedings of the IFHP / CIB-Symposium Berlin, October 14-15, 1991. März 1992

Nr. 17: *Frank, H. P.*
Grenzschichtstruktur in Fronten. März 1992

Nr. 18: *Müller, A.*
Parallelisierung numerischer Verfahren zur Beschreibung von Ausbreitungs- und chemischen Umwandlungsprozessen in der atmosphärischen Grenzschicht. Februar 1996

Nr. 19: *Lenz, C.-J.*
Energieumsetzungen an der Erdoberfläche in gegliedertem Gelände. Juni 1996

Nr. 20: *Schwartz, A.*
Numerische Simulationen zur Massenbilanz chemisch reaktiver Substanzen im mesoskaligen Bereich. November 1996

Nr. 21: *Beheng, K. D.*
Professor Dr. Franz Fiedler zum 60. Geburtstag. Januar 1998

Nr. 22: *Niemann, V.*
Numerische Simulation turbulenter Scherströmungen mit einem Kaskadenmodell. April 1998

Nr. 23: *Koßmann, M.*
Einfluß orographisch induzierter Transportprozesse auf die Struktur der atmosphärischen Grenzschicht und die Verteilung von Spurengasen. April 1998

Nr. 24: *Baldauf, M.*
Die effektive Rauhigkeit über komplexem Gelände – Ein Störungstheoretischer Ansatz. Juni 1998

Nr. 25: *Noppel, H.*
Untersuchung des vertikalen Wärmetransports durch die Hangwindzirkulation auf regionaler Skala. Dezember 1999

Nr. 26: *Kuntze, K.*
Vertikaler Austausch und chemische Umwandlung von Spurenstoffen über topographisch gegliedertem Gelände. Oktober 2001

Nr. 27: *Wilms-Grabe, W.*
Vierdimensionale Datenassimilation als Methode zur Kopplung zweier verschiedenskaliger meteorologischer Modellsysteme. Oktober 2001

Nr. 28: *Grabe, F.*
Simulation der Wechselwirkung zwischen Atmosphäre, Vegetation und Erdoberfläche bei Verwendung unterschiedlicher Parametrisierungsansätze. Januar 2002

Nr. 29: *Riemer, N.*
Numerische Simulationen zur Wirkung des Aerosols auf die troposphärische Chemie und die Sichtweite. Mai 2002

Nr. 30: *Braun, F. J.*
Mesoskalige Modellierung der Bodenhydrologie.
Dezember 2002

Nr. 31: *Kunz, M.*
Simulation von Starkniederschlägen mit langer Andauer über Mittelgebirgen. März 2003

Nr. 32: *Bäumer, D.*
Transport und chemische Umwandlung von Luftschadstoffen im Nahbereich von Autobahnen – numerische Simulationen.
Juni 2003

Nr. 33: *Barthlott, C.*
Kohärente Wirbelstrukturen in der atmosphärischen Grenzschicht. Juni 2003

Nr. 34: *Wieser, A.*
Messung turbulenter Spurengasflüsse vom Flugzeug aus.
Januar 2005

Nr. 35: *Blahak, U.*
Analyse des Extinktionseffektes bei Niederschlagsmessungen mit einem C-Band Radar anhand von Simulation und Messung.
Februar 2005

Nr. 36: *Bertram, I.*
Bestimmung der Wasser- und Eismasse hochreichender konvektiver Wolken anhand von Radardaten, Modellergebnissen und konzeptioneller Betrachtungen. Mai 2005

Nr. 37: *Schmoeckel, J.*
Orographischer Einfluss auf die Strömung abgeleitet aus Sturmschäden im Schwarzwald während des Orkans „Lothar". Mai 2006

Nr. 38: *Schmitt, C.*
Interannual Variability in Antarctic Sea Ice Motion: Interannuelle Variabilität antarktischer Meereis-Drift.
Mai 2006

Nr. 39: *Hasel, M.*
Strukturmerkmale und Modelldarstellung der Konvektion über Mittelgebirgen. Juli 2006

Ab Band 40 erscheinen die Wissenschaftlichen Berichte des Instituts für Meteorologie und Klimaforschung bei KIT Scientific Publishing (ISSN 0179-5619). Die Bände sind unter www.ksp.kit.edu als PDF frei verfügbar oder als Druckausgabe bestellbar.

Nr. 40: *Lux, R.*
Modellsimulationen zur Strömungsverstärkung von orographischen Grundstrukturen bei Sturmsituationen
ISBN 978-3-86644-140-8

Nr. 41: *Straub, W.*
Der Einfluss von Gebirgswellen auf die Initiierung und Entwicklung konvektiver Wolken
ISBN 978-3-86644-226-9

Nr. 42: *Meißner, C.*
High-resolution sensitivity studies with the regional climate model COSMO-CLM
ISBN 978-3-86644-228-3

Nr. 43: *Höpfner, M.*
Charakterisierung polarer stratosphärischer Wolken mittels hochauflösender Infrarotspektroskopie
ISBN 978-3-86644-294-8

Nr. 44: *Rings, J.*
Monitoring the water content evolution of dikes
ISBN 978-3-86644-321-1

Nr. 45: *Riemer, M.*
Außertropische Umwandlung tropischer Wirbelstürme: Einfluss auf das Strömungsmuster in den mittleren Breiten
ISBN 978-3-86644-766-0

Nr. 46: *Anwender, D.*
Extratropical Transition in the Ensemble Prediction System of the ECMWF: Case Studies and Experiments
ISBN 978-3-86644-767-7

Nr. 47: *Rinke, R.*
Parametrisierung des Auswaschens von Aerosolpartikeln durch Niederschlag
ISBN 978-3-86644-768-4

Nr. 48: *Stanelle, T.*
Wechselwirkungen von Mineralstaubpartikeln mit thermo-dynamischen und dynamischen Prozessen in der Atmosphäre über Westafrika
ISBN 978-3-86644-769-1

Nr. 49: *Peters, T.*
Ableitung einer Beziehung zwischen der Radarreflektivität, der Niederschlagsrate und weiteren aus Radardaten abgeleiteten Parametern unter Verwendung von Methoden der multivariaten Statistik
ISBN 978-3-86644-323-5

Nr. 50: *Khodayar Pardo, S.*
High-resolution analysis of the initiation of deep convection forced by boundary-layer processes
ISBN 978-3-86644-770-7

Nr. 51: *Träumner, K.*
Einmischprozesse am Oberrand der konvektiven atmosphärischen Grenzschicht
ISBN 978-3-86644-771-4

Nr. 52: *Schwendike, J.*
Convection in an African Easterly Wave over West Africa and the Eastern Atlantic : A Model Case Study of Hurricane Helene (2006) and its Interaction with the Saharan Air Layer
ISBN 978-3-86644-772-1

Nr. 53: *Lundgren, K.*
Direct Radiative Effects of Sea Salt on the Regional Scale
ISBN 978-3-86644-773-8

Nr. 54: *Sasse, R.*
Analyse des regionalen atmosphärischen Wasserhaushalts unter Verwendung von COSMO-Simulationen und GPS-Beobachtungen
ISBN 978-3-86644-774-5

Nr. 55: *Grenzhäuser, J.*
Entwicklung neuartiger Mess- und Auswertungs-strategien für ein scannendes Wolkenradar und deren Anwendungsbereiche
ISBN 978-3-86644-775-2

Nr. 56: *Grams, C.*
Quantification of the downstream impact of extratropical transition for Typhoon Jangmi and other case studies
ISBN 978-3-86644-776-9

Nr. 57: *Keller, J.*
Diagnosing the Downstream Impact of Extratropical Transition Using Multimodel Operational Ensemble Prediction Systems
ISBN 978-3-86644-984-8

Nr. 58: *Mohr, S.*
Änderung des Gewitter- und Hagelpotentials im Klimawandel
ISBN 978-3-86644-994-7

Nr. 59: *Puskeiler, M.*
Radarbasierte Analyse der Hagelgefährdung in Deutschland
ISBN 978-3-7315-0028-5

Nr. 60: *Zeng, Y.*
Efficient Radar Forward Operator for Operational Data Assimilation within the COSMO-model
ISBN 978-3-7315-0128-2

Nr. 61: *Bangert, M. J.*
Interaction of Aerosol, Clouds, and Radiation on the Regional Scale
ISBN 978-3-7315-0123-7

Nr. 62: *Jerger, D.*
Radar Forward Operator for Verification of Cloud Resolving Simulations within the COSMO Model
ISBN 978-3-7315-0172-5

Nr. 63: *Maurer, V.*
Vorhersagbarkeit konvektiver Niederschläge : Hochauflösende Ensemblesimulationen für Westafrika
ISBN 978-3-7315-0189-3

Nr. 64: *Stawiarski, C.*
Optimizing Dual-Doppler Lidar Measurements of Surface Layer Coherent Structures with Large-Eddy Simulations
ISBN 978-3-7315-0197-8

Nr. 65: *Mahlke, H.*
Mechanismen der Auslösung hochreichender Konvektion im südwestdeutschen Mittelgebirgsraum
ISBN 978-3-7315-0203-6

Nr. 66: *Fosser, G.*
Precipitation statistics from regional climate model at resolutions relevant for soil erosion
ISBN 978-3-7315-0227-2

Nr. 67: *Adler, B.*
Boundary-Layer Processes Producing Mesoscale Water-Vapour Variability over a Mountainous Island
ISBN 978-3-7315-0247-0

Nr. 68: *Kraut, I.*
Separating the Aerosol Effect in Case of a „Medicane"
ISBN 978-3-7315-0405-4

Nr. 69: *Breil, M.*
Einfluss der Boden-Vegetation-Atmosphären Wechselwirkungen auf die dekadische Vorhersagbarkeit des Westafrikanischen Monsuns
ISBN 978-3-7315-0420-7

Nr. 70: *Lott, F. F.*
Wind Systems in the Dead Sea and Footprints in Seismic Records
ISBN 978-3-7315-0596-9

Nr. 71: *Rieger, D.*
Der Einfluss von natürlichem Aerosol auf Wolken über Mitteleuropa
ISBN 978-3-7315-0672-0

Nr. 72: *Loewe, K.*
Arctic mixed-phase clouds. Macro- and microphysical insights with a numerical model
ISBN 978-3-7315-0686-7

Nr. 73: *Piper, D. A.*
Untersuchung der Gewitteraktivität und der relevanten großräumigen Steuerungsmechanismen über Mittel- und Westeuropa
ISBN 978-3-7315-0701-7

Die Bände sind unter www.ksp.kit.edu als PDF frei verfügbar oder als Druckausgabe bestellbar.

Nr. 74: *Metzger, J.*
Wind Systems and Energy Balance in the Dead Sea Valley
ISBN 978-3-7315-0699-7